OKANAGAN ODYSSEY

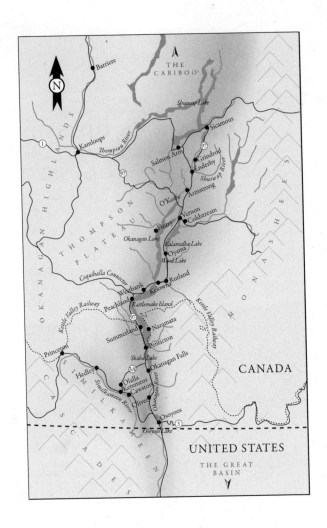

OKANAGAN ODYSSEY

Journeys through Terrain, Terroir and Culture

DON GAYTON

RMB

Victoria Vancouver Calgary

Rocky Mountain Books
www.rmbooks.com

Library and Archives Canada Cataloguing in Publication

Gayton, Don, 1946-
Okanagan odyssey : journeys through terrain, terroir and culture / Don Gayton.
Includes bibliographical references.

ISBN 978-1-897522-81-3

1. Gayton, Don, 1946- —Travel—British Columbia—Okanagan Valley (Region). 2. Okanagan Valley (B.C. : Region)—Description and travel. 3. Viticulture—British Columbia—Okanagan Valley (Region). 4. Biotic communities—British Columbia—Okanagan Valley (Region). I. Title.

FC3845.04G39 2010   917.11'5045   C2009-907204-1

Printed in Canada

Rocky Mountain Books acknowledges the financial support for its publishing program from the Government of Canada through the Canada Book Fund (CBF), Canada Council for the Arts, and the province of British Columbia through the British Columbia Arts Council and the Book Publishing Tax Credit.

BRITISH COLUMBIA ARTS COUNCIL   Canada Council for the Arts   Conseil des Arts du Canada

This book has been printed with FSC-certified, acid-free papers, processed chlorine free and printed with vegetable-based inks.

FSC   Mixed Sources

Contents

I.

The sagebrush was waist high, and dense. I threaded my way silently up the mountainside, between gnarled trunks of the grey and aromatic shrubs. I felt as if I was walking among meditating, grey-bearded ancients. One should show respect and not bump into them, I thought; they could be close to resolving some profound mystery. Despite my efforts, it was difficult to avoid their stiff branches. I can see why ecologists consider sagebrush to be a nurse shrub. Their flaring forms allow for plenty of room underneath for flowers and grasses to reproduce, away from the prying muzzles of cows and deer. I plucked a sprig and put it into my shirt pocket, something I do whenever I am in sage country. The pungent scent of sagebrush speaks to me of natural wisdom, of solitary explorations and of western distance.

I was in British Columbia's Chopaka sagebrush grassland, set in the triangle between the converging rivers of the Okanagan and the Similkameen, hard by the United States border. Black Mountain loomed above me, with its scattered ponderosa pines claiming their niches on rocky knobs. Some trees had been reduced to charred icons, reminders of a wildfire that had swept through 20 years before. I turned back to look south across the rolling sweep of landscape, down into the sculpted valley of the Similkameen and up again to the serrated ranks of the Cascade Mountains on the horizon. The Similkameen River would join the Okanagan just across the border. Together they would run on southward to meet the mother Columbia.

Wind stirred through the sage, carrying complex birdcalls. Together they wove a background tapestry of natural sound, free of the whine of car and highway. A hawk spiralled effortlessly in the thermal overhead. Sun glinted off the needle-and-thread grass, highlighting the subtle greys and blue-greys of the sage. A shallow coulee coursed straight down the mountainside, its bottom a narrow gallery forest of aspen trees. The uncounted quantum of emerald leaves quivered expectantly, as if the whole gallery forest were ready to lift off the mountainside. The Chopaka landscape lay open for me to read, in whatever language I chose, be it of ecology, history or perhaps even the reproductive biology of the copper butterflies that flit about the sage. But the landscape was also open to embrace as a lover, as is and without question. I wanted to fly, to soar with the raptors and look down endlessly upon this and other Okanagan landscapes. I wanted to learn their languages, their emotions and their wisdom.

Valleys are often learned by random encounter: an impulsive hike, a conversation, a local-history book. Even a car breakdown on a back road can infuse bits of clarity onto the local ecology and culture. A well-crafted local wine is also a great asset to this educational process. Having lived in the Okanagan Valley for a few years, I was frustrated by the agonizingly slow growth of my local knowledge. Here was this place, this contradictory Okanagan: part urban big city, part rural orchard and ranch, and part wild ponderosa pine and rockbluff. A local culture that encompasses both Ballet Kelowna and monster truck rallies. I was not satisfied with the conventional view of this region, which is all about real estate, conservative politics and golf. I needed to take the Okanagan on, immerse myself in it, drill down through layers of superficial understanding and traverse it and all its contradictions, from the US border to its northern terminus. So I did that, starting with the Chopaka grassland.

2.

I first experienced the Chopaka on a tour with a park biologist, bumping along its narrow trails in a government truck. It had recently been designated as part of an ambiguous, balkanized entity known as the South Okanagan Grasslands Protected Area. At one point in our tour, we approached the US/Canadian border to look at an old survey milepost. Suddenly an explosion of rapid-fire thunderclaps filled the valley. From out of nowhere a sinister black helicopter appeared, swooped over to us and hovered ominously above the truck. The park biologist instructed me in no uncertain terms to stay in the cab and out of sight. She got out, waved to the helicopter and pointed to her Park Service shoulder patch. The helicopter's occupants, US Border Patrol agents, apparently got the message and veered off. My colleague climbed back into the truck saying, "Gayton, if the guards in the chopper had seen you, with your beard and funky suspenders, we'd be spending all day in some windowless interrogation room." I have always had trouble with borders, but it was disturbing to think that the mere sight of me would provoke official suspicion. However, if the Border folks *had* actually asked me to state the nature of my business on the Chopaka, which involved building local connections between culture and nature and developing my philosophy of honky agnostic land-based mysticism, I would have been detained indefinitely.

The road into Chopaka is long and winding, crossing various cattleguards and jurisdictions from its beginning at Highway Three near Richter Pass to its ending near the border east of Nighthawk. Along the way, it passes through a vaguely defined area known locally as the Kilpoola. The rutted road skirts a couple of shallow ponds and lakes, including world famous – if you are a paleolimnologist – Kilpoola Lake. This lake is the site of a groundbreaking study that determined historical climate change based on the remains of tiny insects buried in the lake bottom sediments. The lab work on that one – sorting through layers of dead insect fragments with a microscope – must surely rank on the Ten Worst Jobs list.

On one trip into Kilpoola, I stopped at one of the smaller ponds to look around and shake off some road dust. Walking along the muddy shoreline, I glanced at an odd scattering of narrow, bright-orange leaves floating near the shore. A closer look set off the shock of recognition: I was looking at the aesthetically pleasing backs of a motionless school of large goldfish. Goldfish! Removed from their original Asian habitat by at least a thousand captive generations, these undignified denizens of nine-dollar aquaria were swimming free – and obviously reproducing – in my Kilpoola! I thought I was ready for anything on my Okanagan odyssey, but I did not expect these placid, bug-eyed fish the colour of orange peels. I was horrified by the presence of these alien invaders, and yet fascinated at the same time. What could have possibly led the original goldfish own-er to drive ten rough kilometres up the Kilpoola road to dump out a nine-dollar aquarium? Was it a tragically misguided sense of compassion? Some twisted form of ecological maliciousness? It couldn't have been a desire to create new fishing opportunities; catching a goldfish would be about as sporty as hooking an abandoned tennis shoe. Or were they surplus baitfish? But then, who would have been fishing in this previously fishless pond? Was I simply witnessing one more example of the infinitely eccentric possibilities of our human in-teraction with Okanagan nature?

Regardless of how they got to the pond, gold-fish are infamous and indiscriminate destroyers of native pond ecosystems. They stir up bottom sed-iments and eat aquatic vegetation. They devour everything, as a matter of fact. My mind raced with potential goldfish eradication strategies. My first thought was author Richard Brautigan's technique: go back to the aquarium store and buy a couple of pet crocodiles. Fight fire with fire.

3.

The Okanagan Valley confounds my typical concentric way of understanding, of incrementally learning about a place from what I know of its nearest neighbour. Points separated by a hundred kilometres *along* the valley axis compare far better than the endpoints of a 15-minute walk *across* the same valley. Moving from a brushy riverbottom along the Okanagan River, eastward, say, up onto a pine-scented foothill, equates to a degree of latitude north in ecological terms and who knows how much in cultural distance. The Chopaka, at the south end of the Okanagan, bears more similarity to the Coldstream grasslands near the northern end than it does to the top of its nearest neighbour, Black Mountain.

The Okanagan River valley lies in a cleft between the Thompson Plateau and the Monashee

Mountains. The river's origin is near the farming town of Armstrong, at the north end of the valley, and it flows due south to meet the Columbia River at Brewster, Washington. Along the way it picks up the Similkameen. Of the river's three hundred kilometres, roughly half is Canadian and the other half is American. The Canadian portion of the Okanagan River system is actually a series of six major lakes – Kalamalka, Wood, Okanagan, Skaha, Vaseux and Osoyoos – with stretches of river in between.

The Okanagan's sister river, the Shuswap, also arises near Armstrong but flows north to join the Thompson, and ultimately, the Fraser River. These two great interior drainage basins – the Fraser and the Columbia – have shared species, waters, ecologies and cultures over time. If you look at a hydrological map of the Kamloops and Salmon Arm area, and throw in random glacial ice plugs or major landslides over geological time, it is not hard to envision a rerouted Thompson River flowing eastward and southward into the Okanagan Valley, which apparently it did for a while.

The Canadian portion of the Okanagan Valley is bracketed by two east–west highways. The Crowsnest Highway, "Number Three," crosses through Osoyoos at the south end. The Trans-Canada, "Number One," crosses near the north end at Sicamous. Highway 97 is the pulsating, north–south artery that links the two. The "97" courses through the bottom of the valley, intersecting most of the Okanagan's population, commerce and tourism. The 97 embraces just as much folklore, funk and heartbreak as the famous American Route 66, but is a better-kept secret.

4.

Wine pairing is a delightful notion. Choosing a Cabernet to go with steak or a Pinot Gris to go with fish is a pleasantly decadent activity. The foodies have definitely invented a great concept, heightening the pleasure derived from both the food and the wine. I liked the idea of extending the reach of my favourite beverage, so I decided to push the pairing envelope.

Wine is a constant companion at my supper table. During the week, I go with carefully select-ed jug wines, moving up to mid-price varietals on weekends. My culinary skills are crude, but after decades of cooking simply to eat, I am now dis-covering the joy in the process. Most of my meals are solitary and I like to read at the supper table – it's kind of like inviting an author to dinner. So I guess it was inevitable that I would eventually create the notion of pairing wine, food and books. Tripling, if you will.

My scientific training demanded that I apply some rigour to the tripling notion, so I poured another glass and created a hypothesis: if chosen carefully, a wine, a meal and a book will synergize to create a memorable sensory experience which can illuminate a local place. In the heat of discovery I thought, why stop there? Given my interest in ecology, why not be daring and add landscape as a fourth element?

I should back up and admit that the origin of the book tripling idea was purely accidental. One evening, I made a beef stroganoff and opened a bottle of robust Shiraz from the Adelaide region of Australia. Giving the rice a few more minutes to cook, I went to the bookshelf to select my evening's reading. My eye caught *The Tree of Man*, a magnificent 1955 novel from the Nobel Prize-winning Australian writer Patrick White, which I had enjoyed reading a few times already. I take no credit for it, but this first tripling was brilliant. In the pages of *The Tree of Man*, I felt I was reading original, unparsed scripts, the very genetic codes for marriage, solitude, war and remembrance. White's characters Stan and Amy taught me a great deal. The Shiraz was equally revelatory and merged well with the rich stroganoff. Lesser novelists than White seek depth in ordinary situations, striving for those hidden meanings behind everyday talk and gesture. White does this instinctively and masterfully. He wrote a big, expansive book, demanding a big, expansive wine. My full-bodied Shiraz was equal to the test. Both the book and the wine draw their energies from the earth and help illuminate our humanity.

Gourmet wine pairings are unappealable; a foodie can say Chardonnay complements Cheez Whiz and it's done. Fait accompli. Every article I have read on wine pairing starts off by saying there are no rules and then proceeds to lay down the rules. My triplings are much less definitive. Each individual finds something slightly different in a book, a meal and a wine. For me, the individual reigns supreme.

I am sure my unorthodox version offends serious gastronomes, who may see pairing as a noble art form not to be toyed with or poached upon. However, I can take shelter under the wonderfully fuzzy concept of wine *terroir* (roughly, "terwar"). The French term refers to a group of wines made from grapes that share the same local ecology, climate and winemaking practices, all of which contribute to give them a specific local personality. Now by extending the notion of local practices only slightly, from making wine to making literature, my heresy becomes legitimate. In a graceful metaphorical leap, I can span the narrow gap between the cultural practices in a local winery to the cultural practice of writing local books. Wine terroir resembles literature in another regard as well: beyond the complex craft of making books or wine is the rich enjoyment of the product, which can be critical or unfocused, depending on the moment.

Like decanting a bottle of wine half an hour before supper, I am opening up the concept of terroir, to let it breathe.

Emboldened by the success of my first tripling trial, I decided to replicate the experiment. The pub in Osoyoos was an ideal laboratory for an additional proof. All in the name of science, of course. It was late afternoon and the pub was busy, but not noisy. I settled into a comfortable booth overlooking the lake and ordered a glass of local Cabernet Sauvignon, a tender variety well suited to the mild winters and hot summers of the Osoyoos area. The wine's colour, lit by sunlight bouncing off the lake, was nothing short of spectacular. Before rubies became gemstones, they must have started life as a Cabernet. The food was creamed local asparagus on toast and the book of choice was my well-thumbed reprint of John Keast Lord's 1866 publication, *The Naturalist in Vancouver Island and British Columbia*. A Victorian-era naturalist, Keast Lord was a veterinarian to the US/Canadian Boundary

Survey party in the 1860s and spent several weeks touring and writing about the Osoyoos area.

Not only do I delight in the frisson of rubbing books and wine together – John Keast Lord and a 2006 Cab Sauv – I can also use them both to help me probe deeply beneath the valley skin. I wanted to learn what I could about the Osoyoos area, from the Anarchist to the east and the Chopaka to the west, and then move on to the next piece of Okanagan geography. I can become a full citizen of this valley through a combination of terroir, curiosity, good wine, local food and good books. Eat, drink, read, look.

After a satisfying tripling session, I left the pleasant confines of the pub and walked down to the public beach. I took my shoes off to revel in the fine sand. In valleys, one's visual field is always bookended; here I had the Anarchist buttress to the east of me and Mount Kruger to the west. It was early spring so the beach was deserted, but soon it would fill up with teenagers, vacationing families and fruit pickers from Quebec. The pickers arrive every cherry season, like gypsies. Mostly in their twenties, they sport backpacks, dreadlocks and bandanas. They like to hang out at this beach in the evenings, drumming, hackysacking, passing the odd joint around and generally having a good time. I wonder how they will fare when the expensive condominium development that looms directly above this beach is completed and occupied.

Osoyoos, as the warmest town in Canada, has its own particular gravity for retirees who have paid their Canadian winter dues back in Brandon or Prince Albert. Osoyoos is their version of Arizona, but with socialized medicine. Some new residents are rudely surprised when the occasional Arctic cold front swoops down the Okanagan Valley and temperatures hit minus 25°c for a few days.

From the beach two conflicting landmarks are visible: Haynes Point to the south of me and Anarchist Mountain to the east. The Point is a long finger of land jutting into the lake. It was named after Judge John Haynes (1831–1888), the original 'get tough on crime' judge, whose favourite form of punishment involved a loop of stout hemp rope. The Anarchist begins with the steep face on the east flank of Osoyoos and works its way up to the summit thirty kilometres beyond. Its name refers to the political philosophies of the area's first postmaster, R.G. Sidley, a civil servant who advocated for no government.

A good friend, Peter Taylor, since passed on, lived in a pleasant log home overlooking Osoyoos Lake. We saw and discussed many things on Peter's rustic wooden deck, wine glasses in hand. I remember nearly dropping mine once as I looked up to see five brown pelicans in majestic flight over the lake. Pelicans are rarities both this far inland and this far north. It was a truly magical sight. Most birds fly outstretched, but pelicans do not; they fly with their long, graceful necks tucked back against their bodies. Their already effortless flight appears even more so, as they fly in such a relaxed position.

Peter and his wife Sue recounted the 2003 Anarchist wildfire, a small but very intense August conflagration that swept up the steep bluff across the lake from their home. They had ringside seats on their deck to watch trees candling, smoke billowing and houses incinerating. Huge water-bomber aircraft skimmed the lake to load up, and then circled upward, engines roaring at full throttle, to get above the fire and drop their payloads in majestic white cascades. Recent migrants from the West Coast, Peter and Sue were absolutely slack-jawed over the raw power of fire in a dry forest.

Peter, a retired English professor from the University of British Columbia, shared an interest in John Keast Lord. Together we sifted through his book, looking for descriptions of local flora and fauna. This is Keast Lord's description of the lake:

If there is an Eden for water-birds, the Osoyoos lakes must surely be that favoured spot. At the upper end, a perfect forest of tall rushes, six feet in height, afford the ducks, grebes, bitterns and a variety of waders, admirable breeding haunts…. The water, alive with fish at all times, is in the summer crowded with salmon.

· Early descriptions of our landscapes fascinate me. I am also irresistibly drawn to the compelling but poorly remunerated field of historical ecology. This is partly due to a quirk in my nature: the more obscure a topic is, the more I am engaged by it. But mainly it is because I sense that past ecosystems can shed light on present and future ones. We don't know how to manage many of our ecosystems because we don't know what we've undone, and historical ecology can fill in those blanks. So I collect early explorers' journals, I haunt the BC Archives photography website and I study tree ring records. I don't yet study insect body parts in lake sediments, but that too may come in time.

Water diversions upstream and dams downstream are slowly choking off the abundant Osoyoos Lake fishery that Keast Lord described. The few hardy salmon that do make it over the twelve (count them!) hydro dams on the Columbia River arrive to a lake with accumulating silt and a narrowing band of suitable water temperatures. As for the perfect forest of rushes, I hope that was not what the errant pelicans were looking for, because it is now a trailer park.

5.

Testalinden, otherwise known as Kilpoola North, is one of those odd, nearby places that are hard to get to. I think Testalinden must have fallen under the spell of the famous Wichita Paradox: no matter which road you take, you can't get to Wichita. A colleague and I headed toward Testalinden through a gravel pit off the Number Three Highway, a few kilometres west of Osoyoos. After following several blind leads, we finally found a track leading into a narrow gulch. Our vehicle quickly became a liability on the rutted track, so we parked and walked on until the gulch opened onto the grassy Testalinden bench overlooking the main Okanagan Valley. A shallow, elongated marsh lay at the centre of the bench. Because the area was relatively flat, someone had of course tried to farm it early on. That enterprise had been brought up short by the brutal rule of "ten inches annual precipitation," below which dry farming becomes risky. You can ranch, but not farm. Osoyoos clocks in right at ten inches (254 millimetres), so there are years you can make a crop, and other years you cannot. At some point the Testalinden grain farming mistake was acknowledged and the site was planted to an introduced crested wheatgrass. This cheap, easy to establish, drought-tolerant pasture grass is also highly invasive, crowding out native plants as it spreads. I muttered and kicked the grass clumps disdainfully as we walked through it.

Our destination was the shallow marsh that had not been farmed up, as it was bound to contain some interesting plants. Shortly after we fanned out along the bulrushes and cattails, my colleague hailed me to come over. He held in his hand the lower half of a lovely broken spearpoint. Complete, it would have been the length of a man's hand. Made from dense, grey-black stone that did not look local, the spearpoint was perfectly symmetrical and notched at the base. The fluted edges were quite sharp. I complimented my colleague on the find and immediately suggested that he donate it to the museum being built on the reserve just outside Osoyoos. Surprisingly, he refused. I argued, but he was adamant about possessing this wonderful object for himself. That event overshadowed the rest of our day at Testalinden, as well as our subsequent relationship.

How many times has this same act of cultural appropriation been repeated across North America, where the White man picks up a First Nations artifact and keeps it, projecting his own desire to be closer to the land – and a way of life – by way of possessing someone else's property?

First Nations pictographs abound in the Okanagan. These images provoke a variety of reactions from the White community, ranging from awe to jealousy. It is not uncommon to see contemporary graffiti scrawled next to and in some cases right on top of the pictographs. The images have attracted non-Aboriginal scholarly interest, some of which also borders on cultural appropriation. A friend of mine who grew up on a ranch near Okanagan Falls had some childhood fun at the expense of a visiting anthropologist. She and a friend went to a small cave they frequently played in, mixed up a batch of powdered charcoal and spit, and drew their own pictographs. Then they breathlessly reported their find to the anthropologist, and led him to the site. He spent the entire day taking detailed notes and photographs. I have always wondered about the outcome of that research.

6.

Not far from Testalinden is the turnoff to Mount Kobau. A stately drive on an amazingly well-maintained Forest Service road brings you up to the broad summit, at nearly 1900 metres. From that elevation you look a kilometre and a half downslope into both the Okanagan and Similkameen valleys. Mount Baldy is visible to the east and the Cascade Mountains to the south. Kobau is a Dark Sky Preserve, treasured by amateur astronomers for its clear night skies and minimal light pollution.

On my first trip up Mount Kobau I inadvertently crashed a star party. Knowing nothing of its astronomical delights, I drove up Kobau late one afternoon to look at the Vasey's sage that grew on its summit. Sage is a curious plant. If you include subspecies as well as species, there are 22 different sages in British Columbia, many found right in the Okanagan–Similkameen. One of them, three-tip sage, is exclusive to these two valleys. So, my trip up Kobau was a form of ecological entertainment: leave Wyoming big sage and three-tip sage habitat at the bottom of the valley, pass through several kilometres of closed forest as I ascend and then pop out into Vasey's sage country on the bald top of the mountain. On this occasion, as I rounded the dusty road's last bend onto the summit, I was confronted by a phalanx of frantic astronomers, all making hysterical downward motions with their hands. At 15 kilometres an hour, I was three times over their speed limit. Behind them, I saw trailers, tents and a technological forest of very, very large amateur telescopes. I came to find out that next to light pollution, the amateur astronomer's worst enemy is road dust. After a stern lecture, the astronomers invited me in for a chat. It turns out that Kobau hosts an extended star party every August, complete with lectures, competitions and awards.

The astronomers say that there is nothing quite like midnight on Kobau, glass of local Pinot Blanc in hand, surrounded by the naked firmament and one's close friends Ariadne, Eridanus and Aldebaran.

Like many regional names, *Okanagan* is geographically ambiguous. Its southern boundary is clearly defined by the US border, even though the valley continues southward, oblivious to nationality, as the Washington State *Okanogan*. Most people would identify the height of land on the Monashees to the east and the Okanagan Highlands to the west as the Okanagan's lateral boundaries. The northern boundary, however, is a bit of a puzzle. Using Highway 97 as a convenient exploration transect, I am clearly still in the Okanagan as I pass northward through Peachland, then Westbank, Kelowna, and on up to Vernon. Okanagan benchmarks of ponderosa pine, patches of open grassland and steep hillsides of light-coloured granite still dominate. Then, just north of Vernon, there is a subtle and immediate change to the vegetation, the landforms and the human settlements. Farms, feedlots and railways now crowd the fertile bottomlands. The valley sides are darker, clothed more in Douglas fir. Aspen, only found in wet areas farther south, is now common across the landscape.

By the time I reach Armstrong, the landscape transformation is more or less complete. Northward through Enderby and Grindrod, I have left the south-flowing Okanagan behind and am now following the north-flowing Shuswap River. The ultimate valley terminus is at Sicamous, on the shores of Shuswap Lake, which drains to the Thompson River and ultimately, to the Fraser. So, Osoyoos and Sicamous sit at opposite ends of what is really a single valley that contains two opposing river systems, two different ecologies, and two subtly different human cultures. The typical Okanagan resident can rattle off five different wine or apple varieties without thinking twice; a Shuswapper is more likely to know the names of five different cheeses or farm implements.

I do like to know the boundaries of whole watersheds, or drainage basins; they are fundamental not only to geography but to the understanding of place. We look toward roads and political boundaries as definers of place, but water is the more significant and enduring layer underneath. Some of my more freethinking colleagues believe that drainage basins should define local and regional jurisdiction lines. It was probably their influence that triggered my desire to know the precise boundary dividing the Okanagan from the Shuswap basin. I knew it had to be somewhere north of Vernon, because I could sense a tiny bevel in the landscape there, an almost imperceptible shift from a faint southerly slope to an even fainter northerly one. I looked up some of the official town elevations, which confirmed my intuitive sense of the bevel: Osoyoos sits at 300 metres above sea level, Vernon at 560 and Sicamous at 350. So, the valley rises steadily toward Vernon and then gently subsides into the Shuswap basin. After poring over hydrological maps, I found the actual boundary between the two watersheds: the Okanagan's Great Divide is just north of Armstrong. Someday I will walk Armstrong's farm fields, to find the two adjacent rivulets running in opposite directions.

In the Okanagan, a rise in precipitation fairly reliably tracks a rise in elevation. So the lowest town, Osoyoos, receives around 300 millimetres of rain and snow water equivalent per year. Penticton (a little higher) gets 330 mm and Vernon (higher yet) receives around 400 mm. But the trend quickly breaks down as you go farther north. Sicamous is a fair bit lower than Vernon, but is on the fringes of the Interior wet belt, so it gets 550 mm. If you ignore elevation and think of Highway 97 as a long, northward gradient of increasing moisture, it begins to make more sense. Osoyoos grows cactus and Sicamous is clothed in cedar.

7.

There is a natural pattern, a shape to rivers: bend and riffle. Gravel bar and cutbank. Main channel, side channel and oxbow. Falls and backwaters. With these devices, rivers dynamically fit themselves to the land. Then the biota fit themselves to the rivers. It is a long, slow dance.

I turn off Highway 97 onto a dirt road just north of Oliver. To reach the river, I first have to negotiate a roof truss yard and then an industrial storage site. Not an auspicious entrance to one of the boldest river restoration projects in North America. Jason Emery is waiting for me at the site. Wearing a stubble beard and thrash metal T-shirt, he looks more like a logger or an off-season fullback than a biologist. "Welcome to ORRI," he says, extending a hand slightly smaller than a dinner plate.

The Okanagan River Restoration Initiative, Jason explains, is recreating salmon spawning habitat that was lost to dams and channelization. Only about 10 per cent of the river portions of the Okanagan remain in semi-natural state, he says; the rest is locked into straight, riprapped channels that prevent floods but provide no spawning habitat. Since straighter also means shorter, the river runs have also lost half of their original length.

We are standing on a large pad of fresh gravel along the edge of the river. I see more fresh gravel and rocks everywhere, along the riverbank and in the river itself. At first, the area looks to me like the site of a bulldozer competition that spiralled out of control. But as Jason points things out, the grand design begins to show through. We are at the downstream end of a newly created side channel, a graceful 200-metre-long arc that accepts about a third of the river's flow at the upper end and then merges it again at the lower. In between the main and side channel is a raw new island. Gravel has been added to the inlet and the mouth of the side channel, creating shallow riffles. The gravel is ideally sized for the few brave sockeye, chinook and steelhead that still make the 1200-kilometre run up the Columbia to spawn here. Here and there in the riffles, strategically placed rocks create slackwater resting pools. Jason then points out a gravel bar on the other side of the river, which I assumed was natural. "Nope, went in last month," he says. "Most of the gravels were taken out of the riverbed when it was channelized in the 1950s. Now we're putting some back."

The original channelization, which was mainly done to reduce flooding, involved dredging out a straight, clean channel and then armouring both edges with riprap dikes. In the ORRI project, the dikes have been relocated away from the riverbed. The flood control remains in place, but now the river has room to play.

ORRI started as the brainchild of local fisheries biologist Chris Bull and hydrologist Robert

Newbury. The Okanagan Nation Alliance (a Tribal Council composed of the eight local Indian Bands), the Provincial Ministry of Environment and hydroelectric authorities in Washington State now also support the project. Years of planning, fundraising and land acquisition went into the making of these meanders. We peer down into the clear water. Ironically, the first fish we see is an introduced bass.

Meander. What a lovely, ambiguous word. A relaxed walk or a slow stretch of water. Or both. A word that embodies both poetry and hydrology. Rivers build meanders naturally; we, on the other hand, require scientists, engineers, politicians, money, planning committees and multiple authorizations. In the 1950s, we retooled nature, with channels, dams, fish ladders and hatcheries. It didn't work. Now scientists and engineers are turning to the ultimate teacher – the river – to learn a new approach. Good for them, for their re-meandering.

8.

I harbour a stubborn resistance to reading docu-
ments on a computer screen, but occasionally
something comes along that is so big (and co-
lourful) that it exceeds my printer cartridge
Scrooge Threshold, and I break down. This hap-
pened with Ted Lea's groundbreaking work en-
titled "Historical ecosystems of the Okanagan
Valley and Lower Similkameen Valley of British
Columbia – pre-European Contact to the
Present." I set up my laptop at the supper table
so I could slowly scroll through the paper. Ted,
whose ecological knowledge of this area runs
deep, had used a recent set of airphotos to map
the various plant communities of the two valleys. Then he dug through the BC government's
voluminous photographic archives until he un-
earthed a dusty set of 1938 airphotos of the same

areas. (As a vintage-airplane buff, I can visualize a Norseman or a Stinson Reliant, or maybe even a Curtiss Jenny, tracking back and forth over the Okanagan landscape.) Comparing the two sets of airphotos, Ted was able to quantify the amount of loss experienced by each plant community due to cultivation, suburbanization, drainage and so forth. Taking that 1938 to 2005 developmental timeslice, he *backcasted* his vegetation data to the 1800s, to the original amounts that existed prior to European settlement.

I had made a recent foray to the Similkameen, so before I settled in to Ted's paper, I laid out a bowl of borscht made of vegetables from Keremeos, some local bread and cheese and a bottle of Cabernet Franc from the Cawston area. In spite of the delightful local food and drink, it was hard to find any joy in this tripling. Water birch and dogwood vegetation has suffered a 92 per cent reduction since the 1800s. Idaho fescue/blue-bunch wheatgrass steppe, the tawny bunchgrass skin of much of the Okanagan: 75 per cent loss.

Antelope brush/needle-and-thread-grass shrub steppe, the rarest of the rare: 68 per cent loss.

I salvaged the evening's tripling by focusing on Ted Lea's achievement. Like the production of a good wine, his work took commitment, patience and rigour.

9.

I have never been a birder, but I do delight in the passionate commitment of our Okanagan birding community. Every spring, they appear, fanning out along marshes, roadsides and hiking trails. Armed with high-quality binos and life lists, these ornithological shock troops patiently observe and report on the highly diverse and incredibly fragile avian life of the Okanagan Valley. Richard (Dick) Cannings is the acknowledged dean of the birding community here. As well as being a celebrated author, he leads tours, does research and monitors his favourite avian group, the owls. Bird counts are a regular feature of a birder's routine, providing raw data for long-term trends. One of the counts Dick coordinates is in a riparian thicket near the south end of Vaseux Lake, between Okanagan Falls and Oliver. Part of the count methodology involves stringing a series of fine-mesh mist nets along a narrow corridor through the thicket. Every fifteen minutes over the course of a day, Dick and his colleagues would carefully disentangle the ensnared birds, record them, weigh them and release them. During one monitoring session the netting snared a surprisingly small number of birds. Further inspection revealed that a local whitetail deer was timing the birders' movements. At ten-minute intervals, when no birders were present, the deer would venture out of his hiding place in the bush and *eat* the ensnared birds right off the net. I was convinced this highly pathological event must be the result of climate change, cell phone transmissions, or maybe jet contrails. It was unthinkable that Bambi would be devouring live, feathered victims, but Dick told me he had seen this carnivory on previous occasions.

The actual Okanagan river reaches, that run between the various lakes, were originally serpentine. Looping oxbows were continuously created, cut off and then recreated by the restless water. The upstream end of each lake was generally a broad delta comprised of a complex mosaic of deciduous trees, dense shrub patches and marsh. This is where the river water would calm, compose itself, and drop its silt before entering the lake. A few pieces of this precious original delta habitat remain; one of them is called the Vaseux Marsh.

One of the more interesting denizens of the Vaseux Marsh is the yellow-breasted chat, a migratory bird of the warbler family. To learn more about chats, I met polymath biologist Jason Emery again. Not one to stand around, he plunged into the dense grass and lumpy ground of the marsh, walking as if he were on smooth pavement. I struggled to keep up, notebook in hand. Jason began to fill me in: "The chat is a beautiful canary-yellow bird with an elaborate songbook, but it's very secretive. You can occasionally hear chats, but you'll rarely see them." Jason and his colleagues estimate there are about one hundred nesting pairs in all of Canada, the bulk of them in the South Okanagan. He continued: "For nesting, they absolutely demand dense, brambly rosebush patches along the river. In fact, we've been able to precisely define their habitat requirements. Dense brush, below four hundred metres in elevation, and less than one hundred metres from water." I do a quick mental calculation. Both Penticton and Keremeos lie right at four hundred metres above sea level, so chats could only utilize the very bottoms of the Okanagan and Similkameen valleys from those two towns southward to the border. Then knock out all the land in the valley bottoms that is less than one hundred metres away from water and that has been converted to roads, subdivisions, golf courses, farmland and communities. In a minute's calculation, I had whittled myself down to practically nothing. I was about to speak, but Jason pre-empted me. "That's right, there's a vanishingly small amount of suitable habitat for these birds." Ted Lea's tables of loss came back to haunt me.

The Vaseux Marsh itself is a prime example of the riparian habitat loss that Ted Lea describes. The river, which used to meander through it, has been straightened and diked. The direct effect of this channelization is the elimination of river and shoreline habitat; the indirect effect is the lowering of the water table in the surrounding flats. As a result, most of the water-loving shrubs and trees are gone. What used to be a patchwork of riparian forest, dense shrubs and cattail marsh now looks like a grassy, unmown football field with a big ditch running through it. In spite of all these constraints, several agencies are trying to rehabilitate the marsh. Jason is a key player in that effort.

Jason goes on to explain another chat threat: the cowbird. Cowbirds do not build their own nests but rather lay their eggs in the nests of other birds, including the chat. The large, fast-growing cowbird chick literally squashes the smaller chat chicks, while the parent chats work themselves to exhaustion feeding the hungry interloper. "That's fiendish," I reply. "Can't we somehow get rid of the invasive cowbird?" Jason chuckles: "Trouble with that strategy is the cowbird is native to North America. It was originally associated with bison on the Great Plains, but its range expanded greatly with the advent of cattle ranching. By turning riparian shrub areas into cow pastures, we've really tilted the balance in favour of the cowbird." I stumble on behind Jason, marvelling at the complexity of the ecological problems we manage to create for ourselves. We arrive at our destination, an experimental planting of native rosebushes and other water-loving shrubs, the first halting steps toward restoring chat habitat. This is Jason's true passion: ecological restoration. He describes the project as plowing up the invasive reed canarygrass, laying down landscape cloth, planting nursery-grown native rosebush stock and setting up temporary irrigation. I look at the three-year-old shrubs. They are surviving, but without a high water table, that is the most one can say about them. To add to the stock of restoration woes, marauding whitetail deer consider the plantings to be their own private, open-air salad bar. In spite of all these immediate restoration challenges, Jason targets what he sees as the root problem: "Virtually every restoration project I know of in this province suffers from the short-term funding syndrome. Governments, funding agencies and politicians all want to see tangible results in a year, maybe eighteen months. Hah. Eighteen years, maybe. You can't deliver an ecosystem in a year and a half."

The south Okanagan and Lower Similkameen valleys graciously host Canada's highest concentration of endangered species. Rare birds, snakes, mosses, toads, badgers, broadleaved plants, insects, sedges, freshwater mussels, lizards, mountain sheep, and fish have all been fleetingly observed in these two narrow, short stretches of valley-bottom geography. Some are not seen anywhere else in the entire country. Most of these species are "peripherally rare." In other words, they are at the very northern edge of their range in the Okanagan–Similkameen and are rare in Canada, but are relatively common to the south of us. To illuminate this curious intersection of nature and nationality, one must indulge in continental bio-geography, a vamped-up phrase that means looking at a couple of big maps. One of the major biomes of North America is called the Great Basin, which covers most of Nevada, extends into eastern Oregon and southern Idaho and reaches northward into the Columbia Basin of eastern Washington. It is an area marked by low elevation, hot, dry summers, relatively mild winters, coarse-textured soils and frequent occurrence of closed drainage basins. Wyoming big sagebrush is both the marker and the icon of the Great Basin.

Zooming down from the continental scale to focus on the Great Basin's range in North Central Washington, one can see a narrowing, north-pointing arm of the biome that follows the Okanagan River. Just south of the Canadian border, the arm splits into the even slenderer Okanagan and Similkameen fingers. So Penticton and Keremeos, respectively, represent

the northern fingertips of one of the largest ecological handprints on the continent. Temperature
largely determines the Great Basin's northern
boundary. To continue the body analogy, if under
current conditions the two Canadian fingers were
to push any farther northward, they would suffer
frostbite.

Occasionally one hears the word desert ap-
plied to the South Okanagan and Similkameen.
While the word has no precise definition, it really does not apply to this landscape that contains
only two diminutive cactus species and plenty of
trees. Local realtors and tourism operators frequently abuse the term, referring to "Canada's
only desert." I'm all in favour of local romance
and pride of place, but for our two fair valleys,
desert is an ecological misnomer.

IO.

My introduction to wine was in the former Yugoslavia. It was both proletarian and place-based. I was just out of high school and my father had a work assignment in the city of Zagreb, in Croatia. Yugoslavia was peaceful in the early 1960s, so he brought the family along. My mother, sister and I made daily expeditions through the city, partly to explore and partly to shop for the evening meal. My mother liked to cook. She outdid herself with the fresh and inexpensive foodstuffs available in Zagreb. We would descend on the cheese shop, the bakery, the butcher shop and the wine shop. At each stop we engaged in often hilarious finger-pointing negotiations in broken English and even more broken Serbo-Croatian. This was far more interesting and satisfying than the shopping routine back home, which consisted of weekly trips to the Safeway. The wine shop we frequented did have a few bottles with labels, but most had none; they simply sat on marked shelves. There were Graševina white wines from the interior, but my parents favoured red wines from the nearby Dalmatian coast. These wines came in big, three-litre returnable bottles, and sold for about 75 cents in Canadian equivalent. I was just old enough to be allowed one glass of the Dobričić or of the Plavac Mali (the presumed ancestor of the famous Zinfandel), to accompany her lavish suppers.

The Dalmatian coast was an hour's drive from Zagreb, so we visited the area on weekends. Dalmatia consisted of a patchwork of small farms and vineyards in rough, hilly terrain. It is of one of the oldest wine-growing regions in the world. The Croats of Dalmatia learned viticulture from their Greek, and later their Roman, occupiers. They went on to develop dozens of varieties of their own, each precisely suited to the limestone and red earths of that dry, coastal hill country.

My connection to wine grew more intimate when I left Yugoslavia to hitchhike randomly through southern Europe. Coincidentally, my destination was always the same place where my ride was headed. Thus, by chance, I found myself in the Rhône Valley of France. Freed from my parents' one-glass limit, I began to understand the powerful romantic dimensions of wine. And as a long-distance hitchhiker, I also discovered the remarkable sustaining power of a loaf of bread, a wheel of cheese and a skin of cheap red wine.

My first Okanagan wine experience occurred shortly after I returned from Europe. A good friend and I went on a camping trip and I brought along a bottle of Côtes du Rhône, which we drank sitting around a campfire on Snehumption Creek. I tried to put the sensation of that dark, earthy wine into words. "It's like a long, beautiful explosion in my mouth," I said. My friend swirled his plastic glass, took a long draught, and replied, "I used to know a girl like that."

In the ensuing years, I have come to understand that wine is place. Wine is celebration. It makes us content with simple foods and grateful for elaborate ones. Wine magnifies the perfume of lilacs after a spring rain, elevates the grace notes in music and writing and helps us realize that every day is an accomplishment – even if we did very little. Wine transforms ordinary confusion into an ecstatic form of wonder. Wine lets us forgive the status quo while it foments us to revolution. Wine reminds us that we are all, fundamentally, romantic poets.

The Okanagan has discovered the wine touring industry, which is now growing rapidly. I believe there is a deeper reason for the spectacular growth of wine touring, beyond the obvious explanations of disposable income, gourmet dining and (alas!) wine snobbery. We live in an increasingly transient, urban and placeless society. Many of us work at movable jobs, shuttling between roughly identical condos and suburbs in roughly identical cities. Wherever we land, we can watch roughly identical television shows and movies. But the human psyche developed during a time when knowledge of, and attachment to, local surroundings was crucial. We still crave place, and wine is our surrogate for it. Wine is the ultimate product of a particular place: influenced by local soil and local climate, vinted locally, bottled locally, sold locally and often drunk locally. We imbibe, and for a moment we feel connected. Wine is a message in a bottle.

II.

The foot trail skirts the western margin of White Lake, a briny and shallow slough tucked in the mountains above Kaleden. I stop for a moment to look across the landscape, which is both magical and mundane. In front of me is the oddly coloured lake water, a muddy yellow tinged with slate green. Surrounding the lake is an empty, boot-sucking mud beach. Beyond lies scrubland and rock outcrop, tall grass and short grass, closed forest and open savannah. Mountains loom on all sides. Most of the small White Lake drainage basin can be seen from this particular spot. Since the basin is nearly closed hydrologically, much of its water stays in place and evaporates, leaving salt deposits behind. The surrounding mountain rim renders the ether of the basin radio-quiet; hence the radio-astronomy observatory nearby.

I think of the White Lake basin as a T-bone made up of three narrow, intersecting valleys. The T-bone's long axis points northwest; that is Park Rill creek's valley. But just before reaching White Lake, Park Rill turns and heads south. Just to the north is Kearns Creek valley. White Lake itself is at the centre of the T-bone. Within the Okanagan–Similkameen region, which hosts a record number of species at risk, the White Lake basin reigns supreme. It boasts a profligate but fragile abundance of birds, plants, mammals, reptiles, amphibians and insects on the edge.

I have walked through the basin's sagebrush behind Dennis St. John, an authority on butterflies, in search of the elusive and endangered Behr's hairstreak. Dennis has moved far beyond simply mapping the occurrence of this diminutive butterfly. As he slowly walks the sage, net in hand, he poses questions to himself about butterfly reproductive strategies, about genetic drift, about butterfly and host plant interactions. Periodically Dennis makes a quick, snake-like jab with his net and I am treated to a brief look at some exotic lepidopteran before he releases it. Dennis and I once sat quietly in the sagebrush watching a male and female Behr's hairstreak perched on the same sage plant, during ideal mating weather. Periodically the male would fly off to chase other butterflies and then return. Yet, the Behr's couple, whose dating profiles seemed a perfect match, ignored each other. After 45 minutes of patient observation, Dennis gave up in disgust. "No wonder they're endangered," was his parting comment.

Even though spring has been dry in the basin, it is late in coming this year. The grassland at my feet, which codes to every nuance of spring, is prospering. Wildflowers, pollinating insects, butterflies and birds are in full riot, or at least to the extent their inherent grassland modesty will allow.

I walk this land often; the White Lake basin is one of my favourite haunts. Every time I come here, I gain an ecological insight or two, but inevitably, these are offset by four or five new ecological questions. Pausing again on the trail, I look up the narrow, rising gunsight that is Park Rill Creek valley. I can just see the top of snow-capped Apex Mountain in the distance. The juxtaposition of late alpine snow and droughty sage in the same viewscape is startling. Nature is offering me a challenge to figure out all the ways that distant white mountain and this nearby yellow slough are connected to each other. And then understand everything in between. The challenge has been there all along, but it is only now that the mountain has brought it into sharp focus.

Here, right in this very basin, I have talked with ranchers, walked with birders, listened to Syilx elders, followed researchers, dug for fossils, monitored grasses, watched painters, identified weeds and studied fire scars. One spring day, I herded a dozing rattlesnake off the sunwarmed blacktop of White Lake Road. I have tried my hand at ecological restoration here and striven to understand the complex layers of local land ownership. I have pored over landscape photos of this valley, both old and new. Over and over, I have tried to analyze what it is about the basin's landscape that attracts me so strongly. From the ground, I have tried to pick out the old glacial lake beach lines that are so obvious in the air-photos. I have tried to imagine the basin clothed in the soaring Metasequoia and the enormous horsetails of the Miocene, whose fossil remnants lie exposed in the roadcuts.

Landscape is the axis around which much of my life seems to slowly rotate. Like many English words, landscape has multiple and simultaneous origins. The suffix *-scape* derives from an old Germanic verb that meant *to create*. That original old German term also morphed into another suffix, *-ship*, which means a condition of being or a position, as in membership, friendship or township. At some point in *scape's* existence, it morphed into the Dutch term *schap*. Following the thread of *schap*, I discover that it combined to form the word *landschap*, which showed up in the late 1500s, referring to a tract of land. As far as I can tell, *landschap* immediately took on cultural baggage when it further evolved to refer to a group of rural people banding together in common purpose, which in this case was to reclaim land from the sea. So this ambiguity, this vagueness about where people end and land begins, bundled itself into the very earliest version of the word landscape.

Equally early on, the word landscape was co-opted by artists and turned into an adjective to describe a new artistic genre called landscape painting, at which the Dutch excelled. Interestingly, it was this painterly definition of landscape which carried over into English, and morphed into the term we use today, which refers to "a tract of land, together with its distinguishing characteristics." In the 1930s, the term landscaping took on an additional meaning when it was first used to describe the arrangement of vegetation for aesthetic reasons. My personal definition for both phases of landscape, the constructed and the natural, is this: "a tract of land, hopelessly entangled in aesthetics and culture."

That snowy Apex Mountain, this crystalline blue morning and the soft grey sage are quietly waiting for me to pull back, to ratchet my narrow focus upward, so I can approach a kind of landscape synthesis. My cell phone is turned off. I contemplate radio silence, the fleeting absence of industrial microwaves and uninterrupted brain circuitry. But I want more than synthesis. I am not content to see and write this landscape through the clear and polished window of knowledge, even synthesized knowledge. I want to see it through the stained glass of creativity.

One of the most commonly observed species in the basin is the naturalist. It is rare to visit White Lake and not find at least one birder. Some of the less common varieties are herpetologists, botanists and entomologists, but you can find them too. Radioastronomers can also be seen in and around the observatory, but their external markings are quite similar to the lepidopterists.

My anti-capitalist hippie roots go all the way back to the 1960s, but in spite of all that, I have a blockbuster business proposal for the White Lake basin. I will erect a modest structure near the junction of the White Lake trailhead and the road. It will be green roofed, straw baled and solar heated. An elegant hand-carved wooden sign over the entrance will announce the Tilley Hat Lounge, which of course will be a non-profit, member-run co-operative. Naturalists of all stripes will be able to gather there after their respective basin field days. Parking will be available, but only for fuel-efficient cars. Once inside, tired patrons can down their telescoping walking sticks, dust off their cargo pants and order a local Pinot Noir or Gewürztraminer. Then they can settle back and brag about their life lists, compare the length of their snakes or swap lies about tiger beetle encounters. A dollar from each glass of wine will go to the White Lake Preservation Fund.

12.

Penticton is an ordinary Canadian city, bounded by two remarkable beaches. The town really comes into its own about mid-July. Depending on whim, you can choose the beach on Skaha Lake to the south or the one on Okanagan Lake to the north. Skaha's beach has coarser sand and is more of a party beach. Its lakefront drive is lined with campgrounds. When the southerlies blow in over the long fetch of Skaha, they hit the shallow beach in spectacular fashion, much to the delight of a crazed cadre of kite surfers. The Okanagan Lake beach is more sedate, with restaurants, motels and elegant homes lining its lakefront drive. The sand at Okanagan beach is fine and the waist-deep shallows seem to stretch on forever. The SS *Sicamous*, a gracious reminder of the age of sternwheelers, lies permanently moored at one end of the beach. Inside the *Sicamous* is a model train layout of the entire Kettle Valley Railway line, which once traversed the tortuous and fractured geography between Hope and Midway.

I swam at Okanagan beach one day during the West Kelowna fires in the summer of 2009. It was 35°c outside and the valley was hazed over with thin smoke, but you could still see Naramata's vineyards off to the northwest. Traffic on Lakeshore Drive crept along, bumper to bumper with lookers, tourists and posers. The beach was crowded with families and teenagers. Out beyond the swimming buoys was an exotic flotilla of paddleboats, kayaks and jet skis. Beyond them, a sleek powerboat towed parasailers back and forth. Every few minutes a Forest Service air tanker would roar overhead, on its way to the fires with another load of retardant from the tanker base at the nearby airport. I paddled along in perfect water doing a slow backstroke, watching the Firecats and c130s as they passed directly above me. It was definitely an Okanagan moment.

A hugely popular summer activity is drifting Penticton's river channel. Every conceivable form of inflatable is used, ranging from Wal-Mart mattresses to smiling children's sea dragons to perpetually surprised inflatable porno dolls. The sun beats down mercilessly on what has to be a dermatologist's worst nightmare. It is an absolutely social event for young people, as groupings form, break up and reform in the placid current. The twenty-somethings seem to heed the warning about drinking lots of liquids, which are carefully stowed in 24-bottle containers underneath their inflatables. For five dollars, you can catch a ride back to the start point via a bus business operated by the Penticton Indian Band.

In late August, Penticton hosts the famous Ironman triathlon. In this event, some 2600 men and women – fit, crazed or both – dive into the morning waters of Okanagan Lake to swim 3.8 kilometres, then bike 180 km over two mountain passes, and then finish up with a 42-km run around Skaha Lake. The transition area is always excessively crowded with spectators, so I like to set up my padded lawn chair in an area where the bike course passes my favourite coffee shop. That way I can sip on a double latte in absolute couch potato comfort as I watch.

I do not quite know how to describe a modern triathlon like the Ironman. It is a combination of fitness achievement, raging insanity and apocalypse. Athletes make supreme physical sacrifices in the name of ... absolutely nothing. A select handful of professionals might win at most a few thousand dollars. Others will run to raise money for a good cause or in memory of a loved one. Some of the tri-addicts do it as a substitute for more destructive addictions. The remaining vast majority do the triathlon because of some personal, and often indefinable, challenge to themselves. By dictionary definition, doing the Ironman is irrational, an activity that is (most certainly!) not play but has no logical purpose. That is what makes the Ironman so fascinating. It is a postmodern endeavour that echoes the very dawn of our species when we ran, chased, ate on the fly, swam rivers and gloried in the physical capacity of our bodies.

The river channel bisects the broad, flat isthmus between Skaha and Okanagan lakes. Penticton lies on the east side, while the Penticton Indian Band land is on the west. The forty hectares of Locatee Lands, now known as Ecommunity Place, are a portion of the Penticton Reserve that the band has been able to maintain in a mostly natural state. Here is one of the last remaining riparian gallery forests in the Okanagan: towering cottonwoods sheltering a dense and brambly understorey of rose and dogwood. Together they host a variety of creatures both rare and common. The elusive yellow-breasted chat nests in Ecommunity Place, and the bizarre, and endangered, tiger salamander favours the cutoff oxbows along the river. The cat-eyed spadefoot toad, the rare western screech owl and the elegantly checkered gopher snake also live on these lands. Richard Armstrong, an elder from the reserve and a Syilx language teacher, is one of the presiding spirits of the place. He regularly takes people of all ages on tours of the area. With his affable and patient manner, I have seen him gradually engage even bored and distracted urban teenagers into the fascinations of riparian nature. Richard has the charisma of a Pied Piper, leading people into deeper connections with the Okanagan landscape.

13.

Summerland is a curious mixture of bedroom suburb, orchard town and retirement community. Originally, the town was located on the lakeshore and transportation was by steamboat. When that era faded into history, the town persisted with packinghouses, cafes and a post office on the lakefront. Slowly, however, the commercial focus began to shift to West Summerland, which sits on a bench two kilometres away from the lake. Sometime after the Second World War, the town fathers decided to move the business community. So now, Summerland has New Town up on the bench and Lower Town, which is now mostly residences and public beaches. Highway 97 goes right between the two. Many tourists assume Summerland is nothing more than the McDonald's, Tim Hortons and 7–Eleven outlets that flank the highway. Many of us local residents are grateful for that misunderstanding.

I have travelled much of Highway 97 and in fact have lived a good part of my life in its asphalt shadow. My first introduction to the Okanagan was along the 97, when as a young punk I worked as a cowboy on ranches near Omak, in Washington State. Then and there I became permanently inoculated by ponderosa pine and bunchgrass. They have been in my blood ever since.

Standing on the highway's noisy shoulders, I am toeing a continuous ribbon of asphalt that runs from Weed, in Northern California, all the way up to Watson Lake, in the southern Yukon. Created in 1953, the 97 is the only highway that carries the same number on both sides of the border. The classic Interior route, the 97 unites the disparate themes of dryness, open space, cheap motels and suburban sprawl. The southern half of 97's 2200-kilometre reach is ranch country, fire country, urban refugee country, sagebrush and ponderosa pine country. From about Williams

Lake northward, short seasons and cooler weather allow dense forests to take over. Occasionally I take my road bike, which was state of the art in 1980, out for a short spin on the highway. If I head north, I can daydream beyond the Okanagan, beyond the Thompson, beyond the Cariboo and all the way up to the Liard country, that lonesome land of black spruce and grizzly bears. If I head south, I can think my way past Omak, where two of my children were born, past the Columbia River, past Klamath Falls and all the way down to volcanoes and sage around Mount Lassen. Then there are a thousand points in between, places where I have stopped, hiked, swum, worked or just lived. Most times I just turn my bike around and pedal the couple of kilometres back home, but it is a comfort to know that my signature highway is there, full of memories, mysteries and possibilities for adventure.

Summerland also hosts a triathlon, a far more modest "sprint" event that is roughly one-eighth of a full Ironman. I typically have a lock on third place in the Men's Over 60 category, except for those years when more than three of us old guys sign up.

Summerland has lots of senior citizens and the little electric mall rider carts are popular with them. You see the carts on the sidewalks and in the streets. Sometimes you see them on the wrong side of the street. Our society's intense relationship with motorized vehicles seems to extend well into the senior demographic. There is a mall rider for everyone's taste. You can get three- or four-wheel versions, enclosed cabs and mag wheels. Depending on your personality, you can choose the Scootie, the Sunrunner, the Rascal or the Marauder. Some are racy, streamlined models, while others mimic the luxury of the Lincoln Continental. Still others are diminutive surreys, right down to the fringe on top. Summerland's version of the vehicle sales lot is the fleet of mall rider carts lined up on the sidewalk in front of the local pharmacy. I do worry about the dying art of walking.

The boundary between town and farm is blurry in Summerland. Orchards, vineyards and pastures finger right into the community. This proximity makes for a certain amount of friction, but the glorious upside is that it connects us more closely to our food and our land. Last Halloween a youngster came to my door dressed as Little Bo Peep, but instead of a sheep, she had an actual pet goat on a leash.

One of Summerland's residential streets that backs right up against an orchard is called Gayton Street. That certainly caught me by surprise, since my surname is quite uncommon and I was never aware of any relatives in the Okanagan. I did a little research in the local museum and discovered the Summerland Gaytons were a prominent orcharding family in the early part of the last century, but none are left now. When I read in the museum's records that the family hailed originally from New Brunswick, then I knew for sure there was a connection, since my people came from Ireland by way of the Maritimes. So, as the johnny-come-lately Summerland Gayton, I am happy to claim the street for my own.

A few Summerland people and places have achieved modest fame: Sam McGee, the inspiration for Robert Service's famous poem, lived here for a time. So did George Ryga, the Canadian playwright and author of the *Ecstasy of Rita Joe*. The agricultural research station, well known for its work on tree fruits, has been located at nearby Trout Creek since 1908. Just behind the station, bordering on the Trout Creek gorge, is the classically elegant Summerland Ornamental Garden.

The Kettle Valley Steam Railway (KVSR) tourist train runs along a small portion of the famous Kettle Valley line. The KVSR's main-squeeze locomotive is a restored 2–8–0 called Number 3716. Built in 1912, it has been around the block, as a working loco in the Crowsnest Pass and later on as a celebrity in the movies *Grey Fox* and *The Journey of Natty Gann*. One of my personal pleasures is to stand on the platform beside 3716 as it departs from the Prairie Valley Station. At first it sits patiently awaiting the command, with its huge black bulk hissing quietly. Then the brass bell on top of the boiler rings out and a single massive *chuff!* of steam issues from the bowels of the loco. Then another *chuff!* – a blast of steam – another *chuff!* and the huge drive shafts tighten against the wheels. Another *chuff!* and now a shudder of movement. Another *chuff!* - the piercing steam whistle – another *chuff!* and the whole KVSR enterprise shudders into forward motion. The *chuffs* of escaping steam are not only music to my ears, there is some kind of masculine eroticism about them. The time between each *chuff* and the previous one decreases exponentially as the train begins to move. I visualize great pistons sliding inside of smooth sleeves. There is no doubt, KVSR locomotive Number 3716 is explicitly orgasmic. The rest of the train, however, is strictly a family show.

Summerland's fruit season begins with cherries. Most are the classic burgundy-coloured Bings, but occasionally Queen Annes are available. Some of the varieties developed at the Summerland Research Station, like Lapin and Staccato, are also beginning to appear. You can get cherries in the two grocery stores, but they are overpriced and the locals rarely buy them there. Fruit stands are where all the cherry action is. You can stop in, buy a plastic bag full for a couple of bucks, chat with the clerk and arrive home with half a bag left. You can also buy directly from one of the larger orchards, some of which offer U-pick at an even lower price. If you poke around Summerland's neighbourhoods you may find a cardboard sign in front of the house that has a couple of massive old trees out back. But the ultimate retail cherry outlet, in my opinion, is the 10-year-old sister and her 8-year-old brother who cleaned off the family cherry tree. Sitting on little green plastic chairs in their driveway behind a pink plastic play table, they have a practically invisible handlettered Cherrys 4 Sale sign taped to the front of the table and three boxes of cherries on top. This is Summerland's version of the kid's lemonade stand, and the 10-year-old girl will most likely have cherry earrings hanging over each ear.

I have planted three cherry trees in my yard: a Bing, a Lapin and a Staccato. They are still spindly seedling trees, but I am getting small crops and making comparisons. The Bing is earliest. In its first full production season, I held off harvesting the tree, waiting for absolute peak ripeness. Then disaster struck. The day before my planned harvest, a flock of murderous starlings moved in and stripped the entire tree, leaving me with nothing but bare pits hanging forlornly from slender stems. So I turned my attentions to the Staccato, which I had previously spurned because its cherries were firmer and not as sweet as the Bings. I realized the cagey breeder of the Staccato had taken the birds into account. He had selected for large, drooping leaves, which completely hide the fruit from the marauding bands of starlings.

The essence of Okanagan fruit is the anticipation. There is the long, dreary spring wait for cherries. One can help pass the time with fresh strawberries and asparagus, but nothing compares to ripe cherries. I delight in their fleshy, overtly sexual succulence; their messy juices and the slick pits that can fly great distances when squeezed between thumb and forefinger. I gorge on fresh cherries for about three weeks until the flavour starts to fail in the late crop. Then I lose interest. But for that three weeks my stomach is like that of a hungry bear during Saskatoon season; it has to adjust to a steady diet of one fruit.

After the cherry crop, there is a lull. In late July, the apricots come on. Shortly after that, the peaches are ripe. One of the joys of peaches is to get a bunch of overripe ones at half price, mash them up with some vanilla ice cream and pour a generous dollop of Okanagan icewine on top. After the peaches, there is another lull and then the table grapes come on. Finally, in the fall, we move into the glorious bounty of the apple and pear season.

I wanted an orchard theme for my next tripling, so I chose Harold Rhenisch's book *Out of the Interior*. Rhenisch grew up in the Lower Similkameen Valley, son of a hardscrabble orchardist. Rhenisch's view of the Okanagan-Similkameen is coloured by those endless hours working in the orchard as a young man. To him, these are valleys of air and of light. I have always liked Rhenisch's writing. He is a poet at heart who never tailors his creativity to a particular audience. The wine for this tripling was a Pinot Gris from Naramata, a funky little community on the east shore of Okanagan Lake. To round out the tripling, I added some artisan cheese and French bread, organic carrots and a plate of sliced Ambrosia apples – a variety developed in the Similkameen. The chilled Pinot Gris was subtle and lively at the same time. Somehow it made the entire meal taste like the first bite. Rhenisch's short book chapters flew by. As they did, I found myself standing with him in that Cawston orchard, in the dancing air.

Life in the Similkameen is a little different than in the Okanagan. From Cawston to Princeton, Similkameeners look practically straight up to see the tops of their mountains, whereas ours in the Okanagan are much more subdued. The Similkameen River, which dominates the floor of the valley, runs mostly wild. I delight in these subtle differences, the ones that distinguish Keremeos from, say, Summerland. Beyond the topography and ecology, there are differences in the human terroir. We read different newspapers, and different issues are top of mind. Summerland has a fair number of fruit stands, but smaller Keremeos is definitely the Queen City of Fruit Stands. Each community differs slightly on the scales of political leanings, economic growth and sheer rural cantankerousness. Perhaps one of the writer's duties is to discover these and other local differences, amplify them and tell their stories. A good wine greatly aids this delicate process.

14.

I do not live next to Okanagan Lake, because I can't afford to. However, I do have the privilege of living very near to this, the queen of the Okanagan water bodies. At 135 kilometres long, it stretches all the way from Penticton to Vernon. The lake gains its water mainly from snowmelt rather than glacier melt, so normal adults (as opposed to children, whose undeveloped nerves obviously lack cold sensors) can swim without fear of cardiac arrest. However, because of the lake's current and its depth, which averages 75 metres and goes as deep as 230 m, the water still has a refreshing bite even in high August. I have always loved swimming outdoors, so those first few early season swims in Okanagan Lake are positively sensuous after a long winter. Some years, with a full-body wetsuit and skindiving bonnet, I am in the water by April. After the initial shock, the cold, clear water welcomes me and I merge with it. Drifting forward, I let the welcome cold slowly seep into my aging joints. Every nerve ending experiences loving assault by both cold and sunlight. I close my eyes, face the sun directly and let the sparkling light penetrate the thin veil of my eyelids.

Okanagan Lake is also my personal fitness centre. It takes a supreme act of will on my part to go out for a jog, somewhat less so to do a bike ride, but absolutely none at all to go for a swim in the lake. I try to maintain some kind of correct swimming form with eyes forward, but the clear view of the lake bottom is fascinating. I usually wind up in a skindiver's stance, paddling slowly and looking straight down. Through my swim goggles I see a huge variety of objects: freshwater clams, suckerfish, sunglasses, waterlogged stumps, old dock pilings, pre-deposit-era beer bottles, engine blocks, steep dropoffs and ever-present golf balls.

After a good dose of looking down, I rotate onto my back. Kicking lazily, I contemplate the valley sides. The east flank of the lake is the most interesting. From Penticton to Naramata it is a long, straight, forested mountainside, with cream-coloured silt cliffs framing the shoreline. Vineyards and wineries are shoehorned into a narrow band between the cliffs and the mountains. North of Naramata, the silt cliffs disappear and the bedrock framework of the mountain flank begins to expose itself. The flank then dips down to accommodate Chute Creek, which some people think actually lies along a geological fault line.

North of Chute Creek, the mountainside shoulders directly out of the water at a 20-degree angle. It is a jumbled, tilted plate of ancient, exhausted and nearly bare rock. Deep fissures and scars crease its face. The mountain's sparse clothing of ponderosa pine burned away completely during the 2003 Okanagan Mountain Park Fire, one of the more severe in memory. As catastrophic as that August wildfire was, it was insignificant compared to what the plate has experienced over geological time. This random quilt of mountain-side displays histories of tectonic collision, volcanic eruption, glacial override and torrential erosion. Like an ancient D-Day veteran, the mountain plate rests quietly and reflectively, displaying the scars and badges of its many battles. I float in the lovely water and try, in my own puny way, to emulate the timeless reflection of that mountain. More wars are yet to come, though, for both of us.

The mountains along the east side of Okanagan Lake abandon their north–south course somewhere opposite Peachland, where they veer dramatically westward. Then, equally abruptly, they switch back to a northward course. The apex of that westward bend is Squally Point. Just beyond lies the prophetic and history-laden Rattlesnake Island. Beneath it is the lair of the legendary lake monster Ogopogo.

If I swim at Crescent Beach, at the north end of Summerland, I can see past the lake's big bend at Peachland to the faint scar of the Coquihalla connector highway as it climbs out of the valley toward Merritt. This artery is the fast lane to the urban delights of Vancouver. If you are not in a hurry to get to the coast or if you are dodging a winter snowstorm, you take the slower, lower-elevation Number Three highway instead.

Antlers Beach, near Peachland, is on a small alluvial fan that juts out into the lake and faces almost directly southward. Like Skaha Beach, Antlers Beach can get pretty exciting when the southerlies blow. Local rumours have it that during the really intense southerlies, surfers appear from out of nowhere.

I live on one side of Okanagan Lake and I look to the opposite side, which means I speculate far more about the geology of that distant side than of my own side. When I do an occasional swim on the Naramata side, I see the drama of my own community's landforms. Giants Head and the Rattlesnake Mountains look as if they were on the business edge of an ancient tectonic plate. During some continent-building crash, they were broken off, tilted downward and then shoved up into the air, leaving a deep trench along their western base. Glaciers then filled the trench with all manner of rock, gravel, sand and silt, and then flattened it, creating a bench for Summerland to sit on. A bench bookended by two mountains.

Trout Creek is a major side drainage of the Okanagan system, arising in the highlands and flowing westward to join Okanagan Lake just south of Summerland. The towering bluffs lining the lake on either side of the Trout Creek delta are fascinating enigmas. Composed entirely of fine, cream-coloured silt, they were laid down at the end of the Ice Age by temporary proglacial lakes that lapped the sides of the main valley glacier. To visualize a proglacial lake, think of the valley as a long, narrow, slightly tilted casserole dish with sloping sides. Then fill it with water, let it freeze, and that becomes the glacier. Then warm it up slowly and spaces appear between the glacier and the valley sides. Pour muddy glacial meltwater over the top of the glacier. Do that long enough until the V-shaped space between the glacier and the valley side fills up with fine, tightly packed sediment. Then continue the warming until the glacier and the proglacial lake are both gone. Now admire the unstable cliffs and old beach lines that are left.

The ultimate disappearance of the glacier at Trout Creek left a 30-métre-high vertical cliff face of highly erodible silt. Right away, the cliffs indeed began to erode, creating strange hoodoos and incising six deep, dry coulees into the land-scape. As a new resident making frequent car trips into Penticton, I would slow down each time I passed the coulee mouths and speculate on what I might find there. After months of procrastination, I finally set aside some time to explore them.

Each coulee mouth is perched several metres above the level of the lake. Entry to them from the busy highway is a brief nightmare composed of the frightening rush of cars and semis, col-lapsing silt and spiny Russian olive trees. Just a few metres in, though, an eerie and windless calm prevails. The coulee sides run steeply upward in a V-shape, until they reach the bases of the vertical silt cliffs. The tops of these cliffs are rain-eroded and wind-carved into fantastical shapes. The south-facing coulee slope is all sage and bunchgrass; its north-facing mate is all pines and Saskatoons. Along the bottom are giant wild ryes and tangled dogwoods. The biggest of the six coulees opens into a rounded amphitheatre worthy of a select audience and some great earth-based play. Flowing aquifers and loose silt have left treacherous sinkholes along the bottom of the coulees, some completely covered over by vegetation. Swallows work the cliff face, magpies cruise overhead and songbirds flit through the dogwood. Thirty metres above me are contemporary horse pastures, orchards and suburbia, but down here, time stretches back to the Pleistocene.

My drives between Summerland and Penticton are now much more relaxed, since I now know the coulees. I do stop to visit one or the other periodically, to get into an unexplored side branch or just to botanize. In all my trips into the coulees, I have never seen anyone else, which allows me to indulge my childhood fascination for secret places.

15.

To really understand wine, I knew I had to do more than just drink it. So I decided to grow some wine grapes myself. I wanted to grow the grapes organically, not from any hardcore commitment, but rather from an interest in the whole process. Organic status meant the usual vineyard inputs such as chemical fertilizers, weed sprays, insecticides, black plastic and pressure-treated trellis posts were all off limits. So I turned to George, my viticultural Jedi master. He and his wife Gerri have created a charmed combination of bed and breakfast and vineyard in the Garnet Valley, north of Summerland. With his particularly Latvian combination of practicality and erudition, George spends endless hours in his vineyard, pruning vines and keenly observing how his nine different varieties respond. Over many glasses of his own product, George and I talked at length about varieties, their adaptations, the wines they produce, and what would be right for the microclimate of my Summerland backyard. Wine grape varieties are intimately tied to the nuances of climate. With other agricultural crops, we have selected for broadly adapted, "one size fits all" varieties, but wine grapes are a notable exception to that rule. We finally settled on Zweigeltrebe, an excellent but little-known Austrian red. I had previously tasted a Zweigelt from Czechoslovakia, as well as local ones from Peachland and Oliver. Zweigelt wines are big and hairy, like me, so of course I liked them.

On a brisk March day, George and I cut sixty canes from the Zweigelt vines in his vineyard, selecting wood that was around pencil thickness, as that calibre is what gives the best rooting success. Back at home, I stood the foot-long cuttings upright in a homemade incubator bed containing a heating tape buried in a thick layer of horticultural perlite. George had shown me how to cut the base of each cane just below a node, at a steep angle to maximize surface area. Then I dusted the cut bases with a rooting hormone. George explained that the trick to rooting cuttings was to keep the base of the cutting warm and wet to encourage rooting, but to keep the top cold. That seemed counterintuitive. Why wouldn't you keep both the base and the top warm, I asked. George, who has an uncanny ability to think like a vine, patiently explained that keeping the tops cool prevents the leaf buds from breaking. If the leaf buds start growing before the roots are established, all your canes will wind up in the compost bin.

After a few weeks of watching and watering my incubator, 40 of the canes took, showing fragile white roots emerging from the cut stem bases. I gently potted them out into four-inch pots with a mixture of peat and garden soil. Then, over the next few weeks, I gradually introduced the canes to the microclimate of my backyard until they were ready to go in the ground.

To dig a hole in my Summerland yard is to experience glacial geomorphology in reverse. At the surface is a layer of silt from that period after the retreat of the glacier, but before any vegetation was established. Fierce post-glacial windstorms moved, sifted and sorted the soil, bestowing on my future yard a fertile but very thin layer of wind-deposited silt. Directly below the silt layer is a veritable minefield of rounded, tightly packed rocks. These were shaped, distributed and then wedged together by raging torrents of meltwater coming off a massive glacier that once filled the entire Okanagan Valley. Below that layer, I presume, will be unsorted glacial till. I don't know that for sure, though, because I've never been able to get beyond the rock layer. The quantity of rock I produce in the course of gardening and landscaping is prodigious. It all goes into a central pile in the backyard, which I refer to as the Great Pyramid of Giza.

About a kilometre away from my yard is an ancient pluton, a great lump of durable volcanic rock called Giants Head. I put the name down to early British colonist prudery, since the pluton looks nothing like a head, but perfectly emulates a woman's breast, especially if viewed from the south. Grape growers value volcanic soil for its rich mineral content, and there are a few pockets of it around the base of Giants Head. For my own vineyard, I have to content myself with the notion that the wind-deposited topsoil might contain a few of the precious mineral elements of that long-ago volcano, plucked off and ground to dust by the glacier.

I had long, meditative blocks of time to think about glaciers as I hand-dug my 40 holes. It is remarkable that such a dominant shaper of the Okanagan landscape is now completely gone. Not in the highest reaches of the Valley's watershed does even a pathetic scrap of that glacier remain.

As I dug, I was astounded to find that a fundamental law of physics was being violated: each completed hole was somehow able to produce twice its volume in rocks. Surely, I thought, there must be a corollary to the Conservation of Matter law, which states that a hole of x volume can only produce a volume of stones equal to or less than x. Apparently not. Holes are normally dug with shovels. My shovel became the glamour tool, for the most part standing around doing nothing. I did the vast bulk of my digging with a six-foot steel pry bar. I read somewhere that in the really rocky wine-growing areas in Europe, farmers will set off an explosive charge in each planting hole, to fracture the subterranean rock and allow the roots to grow through it. I like that idea and I'm sure there must be such a thing as organic dynamite.

The digging did not end with the 40 vine holes. Next, I had to dig even deeper holes for the trellis posts. In the course of digging, my dog would inspect each hole carefully. There must have been famine in his ancestry, as he immediately buries every bone he gets. More than once I saw him standing over a vine or a trellis-post hole, bone in mouth, ready to drop it in. But something in his small dachshund brain would tell him no, these holes are too deep, I'll never get this bone back and I might starve.

Following the organic precepts, I planted each of my precious vines with a couple of handfuls of homegrown compost, some bone meal, mycorrhizal inoculant and a starter of dilute liquid fish fertilizer. Then I spit in each hole for good luck. But all this was just to launch the fragile new vines. If things went well, they would soon outgrow all my hopeful organic additives and begin exploring the native soil profile.

I didn't really think of my grape planting as a vineyard, but right away neighbours and friends began demanding a name. Awfully presumptuous, I thought, since it is only 40 vines. Besides, all the good names, like Dirty Laundry and Stoneboat, were already taken. But the demands persisted. I love the Spanish language, my vineyard is at the far end of our long driveway, and I do like western cowboy lore. So I combined these to come up with a moniker for the vineyard: Yippee Calle.

Before my Zweigelt vines really had a chance to sample Okanagan geology, they got a taste of a very old-fashioned Okanagan winter. The 2008–2009 winter season was one of the toughest in recent memory, with heavy snowfalls at low elevations and an extended cold period. Minus 23°c is a commonly used benchmark for wine-grape bud damage and winterkill; for two nights in December 2008, nighttime temperatures dropped to minus 27°c in my yard. For weeks that next spring the Yippee Calle vineyard became an intensive care unit. As head nurse (actually a candystriper, since I knew so little), I checked the vines twice a day and charted their agonizingly slow recovery. In the end I lost four vines, but George had anticipated the loss and gave me some more canes to grow out as replacements. I was lucky, unlike a number of vineyards up and down the Okanagan that suffered major damage that winter.

The upside of that miserable winter was the following growing season. The Okanagan's 2009 season yielded a small crop of very high quality grapes due to the abundant heat and sunshine through July, August and September. Beach weather is fire weather, but it is also wine-grape weather, and 2009 will be an Okanagan vintage to remember.

Somehow, I take heart from winter cold snaps, summer dry spells and forest wildfires. If these events are part of the cycle of life here, if they help define the Okanagan, then I will plunge deeply into them, experiencing and recording their every nuance.

Wine grapes take three years from cutting to crop. Add another year for vinting and aging the wine. Viticulture teaches many things, not the least of which is patience. But well-tended grape-vines can produce for a century or more, so the rewards are ample. And fruit-forward, with a hint of tobacco.

I have two years of waiting before I can experience the liquid terroir of my backyard. When I do, hopefully it will be a marvellous mix informed by a grape from Eastern Europe, channelled through glacial wind, water and ice, and harmonized by the ministrations of an enthusiastic and barely competent viticulturalist. I think I will farm out the making of my wine to George or one of the many home winemakers in the area. I do not have the equipment, and certainly not the skills. George scoffs at this. "People have been making wine for three thousand years," he says, "how hard can it be?"

As a wine-drinking ecologist, I suffer from a great moral dilemma. I do love the fermented juice of the grape, but I do also passionately love the Okanagan grasslands, shrublands and dry forests that are increasingly being plowed up for vineyards. Often I drink the cheap and reliable Italian Sangioveses, with the hopeful notion that by doing so I might spare a hectare of our native antelope brush or bluebunch wheatgrass. In the end, though, nature and viticulture must come to terms in this valley, because they both form part of our Okanagan identity.

I have another wine mentor in John Vielvoye of Rutland, who is retired from three decades as the province's grape specialist. There is not much, nor are there many growers, that John does not know. He and several other scientists participated in the creation of the massive 1987 *Atlas of Suitable Grape Growing Areas of the Okanagan and Similkameen*. The *Atlas* was an unprecedented synthesis of temperature, precipitation, hours of sunlight, frost-free days, soil textures, aspect and elevation. Ironically, the *Atlas* pointed out that the largest block of Class 1 grape land lies sequestered underneath the concrete and asphalt of the City of Kelowna.

16.

Sharing passions always enhances the experience. George and Gerri shared their passion for grapes and wine with me. In turn, I shared my passion for local ecology, inviting the two of them to an illustrated evening lecture on the Okanagan's endangered animals. The talk featured pictures of the animals and recordings of their various calls. A few weeks later, I received an urgent call from Gerri. "I think we have one of those endangered amphibians in our front yard," she said. "He's living in our ornamental pond and he croaks all night and drives our bed and breakfast guests crazy." I thought for a minute, and decided we needed to be sure. I gave her the number of the Frogwatch hotline, which plays recordings of various amphibian calls, to help identify them (how cool is that!). The next day she promptly confirmed to me that the nocturnal serenader was indeed a spadefoot toad.

The image persists of Gerri standing in the dark next to the pond, cell phone in hand. I can just see her listening to the choices with one ear, and to the pond with the other. "Press one for the leopard frog, two for the pacific tree frog, and three for the spadefoot toad." Meanwhile George and the assembled B&B guests stand by, each holding a glass of crisp, slightly chilled Pinot Gris, breathlessly awaiting the verdict. Once the amphibian's identity was known, as one of the rarest in Canada, the guests were absolutely respectful. Everyone woke up late, to make up for the sleep lost to the spadefoot's nocturnal racket. Spadefoot toads belong to a unique group of night-hunting animals that have vertical pupils. Cats, crocodiles, certain lizards and a few snakes also possess that distinctive pupil, which is more effective than a round pupil for protecting highly sensitive night-hunting eyes from bright daylight.

17.

Pruning is really the central mystery of the entire winemaking process. There are literally dozens of pruning styles, each with local variants. George brought me along on an early spring training session in his vineyard. That first pruning, usually done in February and March, really sets the plant's growth pattern for the year. At the beginning of the pruning session, each vine simply looked like a tangled mess of canes. After a couple of days under George's expert tutelage I began to see the vines as living entities with certain requirements and growth patterns, but also with a certain flexibility.

George's teaching method was heavily Socratic; he would move to a new vine, describe its situation, and then ask me for a pruning scheme before he touched it with the pruners. Pruning is scary for a beginner; it is months before you see the results and you cannot correct a mistake.

George explained how pruning is finding the best possible compromise between the plant's needs and the grower's needs. With the grape's tendency toward lush, almost weedy growth, unpruned vines run wild and produce incredible numbers of shoots and leaves, but little fruit. "Pruning puts the vine in harness and gives it something to do," says George. His grape shears are like an extension of his hand and his pruning routine is methodical. Move to the next vine, stop. Study its configuration. Think like a vine. Think about the upcoming growing season, the next winter and the growing season after that. Then snip, snip, snip. Surplus canes fall away and the chosen ones remain. Then move to the next vine. Art and science.

Grape vines are actually quite weedy, as anyone who has seen an abandoned vineyard can attest. All vines have the same life strategy: grow faster than your neighbours, use them for support and smother them if you can. We have been able to harness this rampant wildness of the vine, train it toward the sun and turn it to producing fine grapes. Sunshine, which Okanagan growing seasons are blessed with in abundance, is a key viticultural ingredient. Galileo mused about how busy the sun was, dragging all those planets through the heavens and yet still finding the time to ripen grapes.

18.

Like Summerland, the nearby town of Peachland started its life as a lakefront steamboat stop. When the steamboat era faded, Peachland's downtown stayed where it was, but the town's focus shifted westward to a massive mountainside suburb of newer, upscale homes and condominia. Like many Okanagan communities, Peachland's economic activity has cycled through ranching, steamboat and rail transportation, orcharding, tourism and now suburban housing development.

Peachland harbours a development story of another kind, a bizarre tale totally out of keeping with this quiet bedroom community. Mohammed (Eddy) Haymour immigrated to Canada from Lebanon in the 1950s. True to his home country's long mercantile traditions, young Eddy combined bold initiative with great entrepreneurial spirit. Starting out as a barber, he built a whole series of enterprises, most of them successful, in Alberta and later in BC. He and his family were eating ice cream in Peachland one day, when Eddy spotted Rattlesnake (previously known as Ogopogo) Island, across the lake. An outlandish vision of a pan-Arab theme park sprang into his head, complete with minarets, pyramids, camels, water taxis and ice cream. He bought the two-hectare island in 1971, and set about realizing his vision. On the Peachland side, he built an Arab-themed motel/restaurant, to accommodate the guests coming to visit the day-use theme park. Then, on the island he erected a pyramid, a mina-ret and a concrete camel.

The next project for the island theme park was to be a miniature Taj Mahal, but it was never built. An administrative battle began, concerning various public health and safety regulations, including the distance of public toilets from the island shoreline. Eddy proved to be a formida-ble adversary. He was not shy about lining up support from powerful friends and public figures. The battle dragged on. Then Haymour did a most incredible thing: he sent fake letter bombs to several government officials. No one really knows why he did that. Perhaps it was from intense frustration with the administrative barriers; perhaps it was at the urging of an RCMP informant; or perhaps it was from a streak of mental illness. Regardless, Eddy had crossed the line and was thrown into Oakalla prison. After months of being held without trial, Eddy was finally charged with possession of brass knuckles and committed to Riverview Mental Hospital. Ever the entrepreneur, Haymour worked diligently with fellow Riverview patients, building and selling handcrafts to the staff. After more than a year, Eddy was released under the condition that he be deported back to Lebanon.

Returning to war-torn Beirut, the injustices of Rattlesnake Island still smouldered in Eddy's mind. In 1976, he and four cousins stormed the Canadian Embassy with machine guns and took the ambassador and his staff hostage. Frantic negotiations with the Canadian government ensued, and after fourteen hours, Eddy and his gang agreed to lay down arms. Remarkably, Eddy was allowed to return to Canada and was given a cash settlement by the Federal Government. The island returned to Crown ownership, and is now part of Okanagan Mountain Provincial Park. All that is left on Rattlesnake Island now are a few crumbling concrete footings.

The debate over Eddy Haymour endures. Was he the victim of provincial politicians who did not like his style and perhaps his ethnicity? Was Haymour just too flamboyant for the conservative Okanagan? Was the man truly mentally ill? Could the fake letterbombs and hostage-taking ever be justified, in light of the injustices committed by the provincial government against him? Was the government in the wrong by belatedly imposing conditions that were not in place at the time of the island's sale? Was it right to let him back into Canada after the embassy attack?

Here was an Okanagan conundrum worthy of an evening tripling. I set about assembling the ingredients: *From Nuthouse to Castle*, by Eddy Haymour himself; *The Trials of Eddy Haymour*, a play in manuscript form written by John Lazarus; and some lamb shish kebabs. For the occasion, I chose a bottle of Muscat Ottonel, an aromatic white wine from Naramata. The ancient Muscat grape is closely associated with Mediterranean and near Eastern countries and it is rumoured to be the source of wines found buried along with King Midas.

As per my experimental tripling conditions, I had already read both books by the time I sat down for the Haymour evening. I flipped through the pages and reflected as I slid the lamb kebabs one by one off their skewers and washed them down with cold, tart Muscat. One of the essential tools for my triplings is a good tea towel, to clean fingers before turning pages.

As I pondered the curious history of Eddy Haymour, it struck me that it was a perfect case study in civics, an outmoded discipline that seems to have disappeared from our school curricula, along with penmanship and deportment. The Haymour saga has all the earmarks of a textbook case. Neither Haymour nor the government were blameless. Individual rights were pitted against collective rights. Both sides invoked the notion of the ends justifying the means. The elusive line between sanity and mental illness came into play. Whether or not xenophobia influenced due process is a cloud that still hangs over the case. The Haymour story is rich, chewy and messy, just like my kebabs and Muscat. Was Eddy guilty, or not? Is Muscat Ottonel a dry wine or a sweet wine?

In spite of the legal and ethical dramas of the case, my thoughts wander back to Rattlesnake Island itself, the unconsidered pawn in this protracted game. No player in any of the lengthy negotiations ever spoke out about the island itself, the only one in Okanagan Lake. No one considered for a moment that in First Nations story, a cave underneath the island was the lair of the legendary Ogopogo, the spirit creature of the lake. No one spoke up for the flora and the fauna. Perhaps now that the island's tumultuous human drama has been put to rest, the nature of Rattlesnake Island can again pick up where it left off.

19.

Westbank always challenges my assumptions. I am self-righteously convinced that rapid growth and suburban sprawl are self-limiting; they sow the seeds of their own collapse. Yet Westbank, a bedroom suburb of Kelowna, has been growing and sprawling for two decades without so much as a hiccup. In between my occasional visits to Westbank, malls full of US-owned franchise stores spring up fully formed, or monolithic big box outlets appear. High-end homes are bolted to impossible Westbank mountainsides. Before they are even finished and occupied, another residential development will begin above them. In South American cities, the steepest slopes are occupied by slums and shantytowns. In Westbank, they are occupied by 5000-square-foot luxury homes.

There are a few bits of native grassland left in Westbank, but for me they evoke nothing but sadness. Abandoned and full of weeds, sporting large signs that proclaim "DEVELOPMENT OPPORTUNITY" or "COMMERCIAL POTENTIAL," these once proud bunchgrass flats patiently await their turn to go under the developer's knife. Their broken-down barbwire fences, relics of some long gone ranch, now hoist the flying colours of development: plastic shopping bags fluttering in the wind. Low-elevation grasslands make up less than 1 per cent of British Columbia's land base. Yet they are forced to carry probably three-quarters of our commercial, industrial, residential and transportation developments, not to mention most of our agriculture. Westbank and Kelowna, towns that are both built on grasslands, now boast fourteen megastores of over 100,000 square feet of retail space, and big-box growth continues apace.

Whatever ephemeral sense of local place Westbank has is rapidly being papered over and homogenized by chain stores that are identical from Savannah to Saskatoon. Local culture is likely to suffer the same fate. However, I must come to terms with Westbank. Suburban sprawl is like a runaway train; no one knows how to stop it. Or, more accurately, we as a society actually *like* riding on this suburban train. We are comfortable with the notion of continuously inflating real estate values as the primary engine of our economy, comfortable with the endless conversion of natural landscapes, comfortable with one-stop, franchise shopping, and comfortable with staggering levels of energy consumption.

Our default urban framework here in the Okanagan is: live in the suburbs, drive the kids to school, drive to work downtown and then drive to the shopping mall on evenings and weekends. This arrangement, created in the 1950s, has proven to be immensely durable and successful. The entire framework is built on a foundation of cheap fossil fuel, a very questionable base but one that has proved to be remarkably durable as well. At least for now.

For my Westbank tripling, I chose a blended red from a winery at the foot of Mount Boucherie, a volcanic remnant that presides over Westbank. For the book, I reached back to a 1961 urban planning classic, *The Death and Life of Great American Cities*, by Jane Jacobs. The meal was as urban as I could think of: taco chips and a fresh salsa made from tomatoes, green peppers and cilantro. Jane Jacobs, in her quiet way, shot an enduring arrow through the heart of North American urban planning. She advocated dense, pedestrian-oriented neighbourhoods, a variety of housing forms, and a blend of commercial, residential, public and light industrial uses in the same area. Jacobs said:

> … make it an onerous trip to drive a car through the main streets of your downtown. Make the sidewalks wide and the streets narrow, the blocks short, the lights long, and never ever switch to one-way streets in the town centre.

Even though *Death and Life* is half a century old, it is still widely read and quoted today. Like the Gamay Noir in my blended wine, her ideas have aged well. Unfortunately, Jacobs's compelling notions of what makes a city vibrant are still revered in theory but almost totally ignored in practice. As I flip pages, scoop salsa and sip red, I mentally review Okanagan towns. Oliver and Vernon, with their dense, centralized downtowns, probably come closest to Jacobs's vision. The least Jacobsean prize would have to go to Westbank, with its commerce strewn for fifteen kilometres and its traffic slowed by a grand total of ten stoplights.

Jane Jacobs was a great fan of urban diversity, and of the synergies that arise when different human groups and city functions are rubbed up against one another. The bold winemaker who made my blended red took Marechal Foch, Merlot and Gamay Noir grapes, and made a resonant wine. He too knew of the value of synergy.

Occasionally, I pass harsh judgment on Okanagan wines as being overpriced compared to comparable quality Italian, Chilean or Australian products. At the same time, I do have to remind myself this is a new industry here. The itinerant Oblate priest Charles John Adolf Felix Marie Pandosy, planted wine grapes in the Kelowna area in 1859, but serious winemaking is barely 20 years old in this valley, making it one of the youngest winegrowing areas in the world. So, I cut them some slack and shell out the extra couple of bucks.

As a wordsmith, I do delight in the outrageously inventive descriptions affixed to the labels of Okanagan wines. "Rumours of blackberry." "Profound depth of character." "Hints of cassis." (What the hell is cassis, anyway?) "Supported by firm leather tannins." "A fruit-forward nose." My hat is off to the anonymous people that write these wine descriptions. They combine literary innovation with tremendous powers of sensory description, all in 200 words or less.

Toward the end of the meal, I slipped another book into the tripling: *Summerland*, by George Ryga, published in 1992. Ryga was blunt about our approach to local and regional culture:

> With the exception of Quebec, where cultural expression has long been integrated into national survival, the rest of Canada, the second richest in the world, is among the poorest in its cultural deprivation. And because of this, we remain ripe as a dumping ground for the commercial cultural refuse of the world. Even worse, we mimic it....

My urban tripling produced a highly entertaining evening, but it resolved nothing. The challenges remain: how do we create liveable Okanagan cities, viable local cultures and sustainable Okanagan landscapes?

20.

The north end of Kelowna is a bustling, car-dominated mix of strip malls, light industry and suburbs. Just beyond the new University of British Columbia Okanagan campus, and surrounded by development, lies a little remnant of Okanagan dry forest. It is not much bigger than a postage stamp and has all the signs of abandonment, plus a diverse collection of urban refuse: a broken lawn chair, a twisted metal stair railing, a decaying plastic flamingo, a carefully tied noose hanging from a pine branch, empty beer cans and plastic shopping bags.

I came to this forest because I am interested in fire history, and its veteran tree stumps have old fire scars on them. I did not want to use a chainsaw to cut the stumps, since the land had suffered enough abuse already. So I borrowed a heritage bucksaw that a logger friend had mounted on his rec room wall. It was five feet long, had raker teeth the width of my thumb and probably hadn't been used since the 1930s. But it was functional. Once I got the hang of making long strokes, the sharp rakers seemed to just flow through the wood. I grunted and sweat like a pig as I sawed, but finally came away with three cross-sectioned "cookies" from three different stumps. Back in my shop, I took a belt sander to the cut surfaces. Starting with coarse sixty grit, I worked my way patiently upward, stepping a hundred grit at a time. As I did so, the rough cross-sections of weathered ponderosa pine began to smooth out. After three hundred grit, the surface was like silk; after five hundred, it felt like glass and glowed like amber. My shop took on the redolent odour of a pine forest after a summer rainstorm. Finishing up with six hundred grit, I could see even the finest tracings of the trees' growth rings emerging, along with the delicate black scorch marks of old fires.

Welcome to the young and vibrant discipline of historical ecology. Fire-scarred tree sections are among the dozens of artifacts that provide clues to the past life of our Okanagan landscapes, and offer guideposts for their troubled future.

These trees were most likely logged in the middle twentieth century. Without actual records, there is no way of knowing if a stump is ten years old or fifty years old. In the 1880s, when the parents and grandparents of these loggers arrived from Europe, there were a few fleeting decades when they wrote, photographed and sketched a virginal valley that as yet they had little impact on. The tree cookies, plus those early accounts, are some of the raw materials of ecological history.

Fire-scarred, or "cat-faced," trees are fairly rare. They are produced when a fire burns away part of the bark, usually near the base, but does not kill the tree. Then subsequent fires leave scars on the exposed wood. My polished tree cookies yielded up the years of the forest fires, showing periodic narrow burn marks in between the annual rings. Because I did not know what year the trees were cut down, I could not fix the fire scars to specific years. I could, however, determine the interval between fires. The first cookie had two scars, seven years apart. The second cookie had three scars, with 34 and 11 years separating them. The third cookie had eight scars, with an average interval of six years. As all three stumps were fairly close to each other, some of the scars are certainly products of the same fire. But I cannot determine that, not without a lot more sampling. The other complicating factor is that not every fire will scar every cat-faced tree, so a degree of speculation is involved.

I set out my three cookies on the dinner table so I could muse over them for a few days. They were like history books, with events recorded in charcoal rather than print. Some of the fires that left scars would have started from lightning strikes. Others were started by First Nations folks, who managed the Okanagan landscapes according to their own canons for thousands of years.

The tree cookies also told me about the drought years that produced absurdly narrow rings, and the glorious wet years when the trees swelled to bursting with new wood.

We do have a curious notion of nature as a static, unchanging entity, always there for us when we turn to it. Nothing could be further from the truth. Nature has more twists and turns than a coiled snake; more bumps and grinds than a mountain creek. There are fires and floods and snow avalanches. Beetle infestations and fungal epidemics. There are blowdowns and ice storms and protracted droughts. Layer in the drama of human impacts on top of the natural disturbances and you have a hugely complex mosaic that historical ecology can help us sort out.

There was a time before these trees, when small lakes like Kilpoola trapped airborne pollen of the prevailing vegetation and held it on their bottoms, in secret, layered muds that only palynologists can decipher. These pollen scholars tell me that our Okanagan vegetation swung vertiginously from spruce to sagebrush and then back again. And right in the middle of those pollen sediments, like a region-wide bolt from the blue, lies a thick layer of ash from the eruption of Oregon's Mount Mazama 7000 years ago. Going back further yet, the Okanagan Valley was glacier from ridgeline to ridgeline. This grinding monstrosity arbitrarily decided which areas got soil and which got stone. Early beaches hang high up on contemporary mountainsides. Back further yet? How about to the Middle Eocene, when ancient and tropical trees left their delicate fossil record around White Lake.

I know this is way too much ecological history to embrace, but the artifacts themselves are compelling works of art in their own right: haunting early photos, fascinating traditional ecological knowledge, elegant tree cross-sections, early explorer accounts, and the spidery tracings of tree leaves trapped in millennial stone. The materials of ecological history can easily become objects of obsession. I once spent several months of my spare time retracing the steps of George Mercer Dawson as he photographed his way down the Kootenay River in 1883. I was able to locate and retake most of his images, and study the landscape changes that have taken place. Dawson's photographs are like secret windows into a nearly forgotten ecological past. As I followed Dawson, I was distracted by the declarative power of his images, cast as they were onto glass plates using a bulky camera the weight of two concrete blocks. I do my contemporary retakes with a slick electronic gadget that weighs less than an apple, but the results pale in comparison.

Someone recently said to me: "Gayton, all this historical ecology stuff is meaningless now because climate change will alter everything, in ways we've never seen before. Climate change means we're starting with a blank slate." I could not disagree more. Nature is about change, and historical ecology can provide us with a record of ten thousand years of it, if we're willing to look. More than just for the art or the personal fascination, we need to improve our abysmally poor understanding of how Okanagan ecosystems work. Historical ecology can provide us with some key blueprints.

Every scrap of land has a past, even that little abandoned forest in the suburbs of Kelowna. Our valley's future is written in the ecological histories we haven't read yet.

21.

In the summer of 2003, Okanagan beaches were crowded. Whole families would decamp to the lakeshore for the day to escape the sweltering heat of the house. By midmorning, all the lakeside parks would be packed with family encampments built from folding lawn chairs, coolers, baby cribs, blankets and umbrellas. The most sought-after locations were those with a grassy lawn, a picnic table under a shade tree and not more than a Frisbee throw to the water. The sunny days were welcomed early on, but slowly they became relentless. Kelowna went 44 consecutive days without a drop of rain, a new record. The surrounding forests became parched and brittle. The open burning ban came unusually early, in mid-May. After two years of below average precipitation, the southern Interior was beginning to fray around the edges. More and more of Okanagan daily life shifted to the beaches and lakeside parks.

Beach weather is also fire weather. Through the early summer, the five-category Fire Danger Rating crept upward. By August 1st, the entire southern Interior was placed in the EXTREME category. Dry lightning storms put everyone on edge. The Penticton Fire Zone crew had their hands full: by mid-August, they had dealt with 95 fires, half of which were still burning. At night, the moon was the colour of blood oranges. During the day, a bronze haze lay over the valley. Some days, a fine dusting of ash would settle out of the sky. One facet of life in the bottom of a valley is that you don't know what is happening, in terms of weather or fire, just beyond either ridgeline. It was a nervous time.

A little after midnight on August 16th, a lightning strike touched down on the rocky, inaccessible mountainside of Okanagan Mountain Park, just a stone's throw from Rattlesnake Island. The resulting fire was spotted at 2 a.m. By daylight, a helicopter was on the scene, bucketing water from the lake. There were no houses anywhere near the fire, but the local fire boss recognized the possibility that the fire could creep eastward across the sparsely timbered mountainside, toward Wildhorse Canyon. If the fire did get into the heavy timber in Wildhorse, it could make a run northeastward toward the southern suburbs of Kelowna. He directed ground crews to start building guards on the east flank, between the fire and the canyon. Then he called in the big yellow CL415 water bombers to hit the same area. The fire appeared to be under control and winds were light. The bombers finished their runs and headed back to the base in Kamloops, to refuel and work on other fires.

Fire is influenced by a complex but predictable set of conditions. In the air, these conditions are temperature, precipitation and relative humidity. On the ground, the amount and type of fuel, the vertical and horizontal distribution of fuel, and fuel moisture content all play a role. The fire itself can be characterized by temperature, flame length and rate of spread. Then there are the ignition sources, both human and natural, and their probabilities of occurrence. Fire scientists and wildland firefighters have merged these factors into several indices that allow them to characterize, communicate and even predict the exact nature of a particular wildfire.

But there is one dominant factor, one unpredictable joker hidden in every fire deck: wind. We know what wind does to a fire. For instance, we have documented that it only takes a ten kilometre an hour wind – barely a light breeze – to start a spot fire moving through a dry ponderosa pine forest. We know a great deal about what wind does to fire; we just cannot predict wind's local occurrence, speed and direction. We know too that intense fires generate their own convective winds, which carry burning embers aloft. Fires in this advanced stage of development are truly like dragons; their own hot breath intensifies the destruction as they sow the surrounding landscape with their fiendish progeny of airborne embers. At about one o'clock in the afternoon on that fateful day of August 16, 2003, the joker showed its face in the form of gusty, 35 km/h winds up Wildhorse Canyon. Fire K50628, soon to become infamous as the Okanagan Mountain Park Fire, ramped up and started spotting ahead of the fireguards. By 2:30 p.m. it had jumped containment and begun its run up Wildhorse Canyon. The rest is history: the fire burned until September 8th and consumed 239 homes in the south suburbs of Kelowna. Thousands of residents were evacuated and millions of dollars were spent. Nature had risen up and provided the valley with a defining event. Days were filled with the sound of aircraft and the smell of smoke. The evening news provided images of candling trees and charred homes. We Okanaganites all remember what we were doing in the late summer of 2003.

22.

Scientific literature has its own fascinations, and the literature of wildfire is no exception. There are a couple of scholarly journals devoted strictly to fire; papers on fire show up periodically in other journals; and a steady trickle of books are produced. There is also the "grey literature" on fire: the consultant's reports and other studies that do not make it into refereed journals. I keep up with this literature, as fire is one of the keys to understanding the Okanagan. So much of a key, in fact, that a wine tripling was in order.

For the evening event, I selected: Stephen Pyne's seminal *Fire, A Brief History*; a couple of journal articles on fire in the Okanagan Valley; a short story by James Marshall called "Part Time Angels" from his collection *Let's Not Let a Little Thing Like the End of the World Come between*

Us; and a contemporary memoir from Melody Hessing called *Up Chute Creek*. The meal consisted of barbecued steak from a local ranch, oven-warmed whole-wheat pita bread, and a salad containing fresh tomatoes from the garden. The wine was a bit of a challenge. I had to canvass many wineries before I found a wine dating back to 2003, the year of the fire. Finally, one of the larger wineries delved into their wine library and came up with a Cabernet for me. I had never heard of a wine library before and was totally intrigued by the concept. I could only imagine the amount of self-restraint required to qualify as a wine librarian.

The Okanagan Valley is known as a fire-maintained ecosystem; its grasslands and dry pine forests historically burned at intervals ranging from roughly five to 35 years. Lightning started some of those fires and First Nations people, who used fire extensively as a land management tool, started others. This frequent-fire regime kept the fuel loads down, thinned the forests to a healthy parkland state, encouraged grasses, shrubs and wildflowers, and kept water and nutrients cycling properly through the landscape. Every fire does some damage, but the paradox of fire-maintained ecosystems – and the Okanagan is just one of many throughout the world – is that the less often they burn, the more damage they do.

The Flea Market at Okanagan Falls is a delightful open-air bazaar offering fancy belt buckles, old LPs, velvet paintings and ornamental china. I think of it as the epicentre of the Okanagan Valley; if you put a colossal pin through at the Flea Market and spun the whole Okanagan, it would not wobble. Looking beyond the tents and awnings of the market, up to both valley sides, you can see ponderosa pine trees dotting the rocky landscape. The hillsides are an archetype of the entire valley. Every year these sparse forests create fuel, in the form of shed needles, dead branches and dead trees. In our dry environment, the biological breakdown of these dead materials is exceedingly slow. In contrast, the fuels of wet coastal forests are quickly broken down and mixed into the soil by an industrious army of fungi, moulds and insects. In dry forests, on the other hand, fuel stays as fuel – often for decades – before it finally becomes non-flammable humus. The primary fuel breakdown mechanism in the dry forests of the Okanagan-Similkameen is fire.

Early European settlers to the British Columbia Interior took a pragmatic fear of fire to a higher level, seeing it in biblical terms. They were convinced fire was Satan's handmaiden and the fact that the non-Christian Okanagan Indians used it was simply further proof. Organized wildfire suppression began in British Columbia in 1912, with the creation of the Forest Service, and kicked into high gear after the Second World War. The Forest Service's Protection Branch became a world leader in the business of locating and extinguishing wildfires. The unintended consequence of this spectacular success was a growing host of low-elevation forests that had missed one, two or even three fire cycles, and were overloaded with fuel.

Smoke from the 2003 wildfires ruined some grape harvests, but others benefited from it. "Smoky" is a commonly used descriptor for red wines, but this Cabernet Sauvignon I had purchased for the tripling had moved beyond analogy into the actual experience of prolonged and dangerous smoke. At six years old, it was probably the oldest wine I had ever tasted. I ate a bit of salad first to cleanse the palate and then took the first draught. The first blush created a large and marvellous open space for the subsequent flavours to play out, which they did. After all, a fire year is a hot year and heat allows wine grapes to produce more of those hundred-odd compounds that contribute to taste. After many seconds in the mouth a certain dryness arose, at the very back end of the finish. Was the wine actually smoked, or just smoky? Who is the better painter, Rembrandt or Van Gogh?

Like wine, smoke contains a host of mysterious components and it comes in endless variety. Smoke also contains an ecological trigger, convincing certain native plant seeds to germinate in order to take advantage of the surplus nutrients and empty growing spaces created by a fire. Researchers proved that connection not only with real smoke, but also with liquid hickory smoke from the condiment aisles of the local grocery store. I certainly have to give credit to the imaginative plant scientist who figured that one out.

Most of us would not think of ourselves as smoke connoisseurs, but the average person can actually identify quite a range. Campfire, wildfire. Burning leaves, burnt toast, burnt meat. Cigarettes, cigars, marijuana, incense. Burnt rubber, burning sage, burning grass, burning garbage. Many people experience wine tastings, but no one has ever been to a smoke smelling. Yet our range of smoke discrimination is impressive. As for woodsmoke, it has been a constant of the human experience since we came down out of the trees.

At a certain point in the Okanagan Valley's development, smoke met suburbia. The spectacular landscapes of the Okanagan became a magnet for the first of several waves of urban refugees beginning in the 1970s. By the 1990s, this back-to-the-land movement had gone seriously upscale and morphed into suburbia. Housing developments pushed into the side valleys and up the hillsides from Osoyoos to Armstrong. "Close to nature, minutes from downtown," became the Okanagan's real estate mantra. Suburban development and total fire suppression slowly aligned themselves on a collision course. As the suburbs expanded, so did the fuel loading. Forests thickened up with a dense carpet of post-suppression trees and woody material accumulated on the ground. Largely ignorant of wildfire and lulled by the effectiveness of the Forest Service's Protection Branch, new homeowners built right up against the edge of the forest. Then they adorned their homes with flammable shake roofs, wooden decks, woodpiles and outbuildings.

The Mount Boucherie Fire of 1992 was probably the first of the Okanagan's modern interface fires. Mount Boucherie, a great lump of volcanic rock overlooking the rapidly developing community of Westbank, burned fiercely in a May wildfire. Fortunately, no homes were lost. Hard on the heels of that fire was the 1994 Garnet Fire, which burned the hillsides just to the east of Penticton, triggering a large evacuation and the loss of 18 homes. Then the summer of 1998 saw the eruption of the Salmon Arm Fire; six thousand citizens evacuated, 40 buildings destroyed. Then came 2003, the year of the Okanagan Mountain Park Fire, as well as the Anarchist Fire, which consumed two homes. Farther north, in Barriere, the McLure Fire burned out 73 homes and a sawmill. Another interface fire year was 2009, with suburban conflagrations near Westbank and Fintry in the Okanagan Valley, plus Goldbridge and Lillooet in the upper Fraser Valley. Of 2009's hundred-odd major fires, 27 triggered suburban evacuation orders.

Fire is as compelling as sex and as incendiary as religion. After scanning the scientific papers scattered across my dinner table, I turned to

Pyne's book. He argues we do ourselves a disservice by shaping and defining fire strictly through the language of physics. The study of fire, he maintains, should be an integrative discipline combining ecology, sociology, history, psychology, *and* physics. Pyne argues that fire is part of what defines us. One of my favorite quotes of his reads:

A grand dialectic emerged between the fire-proneness of the earth's biota and the fire capacity of humans such that they co-evolved, welded by fire to a common destiny.

Not far from my home in Summerland, there is a roadcut that slices down through an inconvenient mound of glacial till. Three metres down from the top of the cut are four thick bands of charcoal, each about a hand's breadth apart. I do not know whether the charcoal is from fires that occurred after the last glaciation or perhaps some previous interglacial period. But no matter, the message of those black bands is still clear: wildfire has been with us for a long time.

I poured another glass of the fateful smoky Cabernet, using the wine carburetor that Gerri gave me. I dearly love this little implement, an oddly shaped aerating funnel that you direct the wine through on its way from bottle to glass. In the upper part of the funnel, an intriguing little vortex is created as you pour. In the lower part, two tiny aerating holes penetrate the sides of the funnel, somehow letting air bubbles in without allowing any spillage. As the wine goes through this apparatus, it makes an outrageous gurgling noise like someone with a bad head cold blowing their nose. Shortly after I got the carburetor, I had a dinner party, and one of the guests was a physicist. As I poured his glass through the carburetor, I seized the opportunity to display my knowledge: "So, Dan, does this carburetor demonstrate the Bernoulli principle or the Venturi principle?" Dan examined it carefully and handed it back. "Neither one," he replied.

At any rate, my carburetor is designed to quickly aerate wine, so it can breathe and develop its aromas. One of my oenologist friends claims that thirty per cent of a wine's taste is actually in its smell. I had a graphic demonstration of the power of aeration on wine when I once spilled a dollop on the supper table. Since there was no one around to be horrified, and since I had paid over ten dollars for the bottle, I bent over and sucked the spill up with my mouth. Wow: a rather demure wine suddenly had bells, whistles, neon lights and gyrating pinwheels of flavour. I am considering a new wine tasting protocol wherein the wine sample is delivered not in a glass, but in a flat saucer.

I don't know the physics of my beloved wine carburetor, but I do know it is a substitute for decanting wine before supper, or pouring a glass and letting it stand for ten minutes, or slurping it straight off the table. What the device amounts to is a time-saving method for slow-food devotees.

I took another long, heavily embroidered draught of the smoked Cabernet and opened

James Marshall's compelling short story. "Part Time Angels" takes place during the Okanagan Mountain Park Fire. In it, a young couple stays on in an evacuated Kelowna neighbourhood. They break into empty homes and choose items they think the homeowners should have saved, but didn't: a dress, a picture, a book. They wrap the mementoes tightly inside plastic bags and submerge them (and themselves) in a backyard swimming pool as the firestorm races through the neighbourhood. Like fire itself, Marshall's story is totally unexpected, random and absolutely compelling.

Wine devotees like to talk about finish. A poor wine with no finish sounds an initial, single note in the mouth, and then it is done. Wines with long finish hit several flavour notes in a sequence that can last as long as seven or eight seconds before the symphony subsides. The amplitude of wine finish is like the arc of a story. I like long finishes and ambitious arcs.

Next, I turn to Melody Hessing's *Up Chute Creek*, which contains a first-hand account of living through the same fire. For days the conflagration burned close to their hand-built and well-loved home on the flanks of Okanagan Mountain, north of Naramata. Melody and her husband Jay endured the terrible dilemmas of the threatened rural homeowner. Shall we stay and defend our home or do we put safety first and evacuate? If we do evacuate, how will we choose which mementoes and treasures to take and which to leave? Hessing's writing conveys the anguish of those decisions. I am sure the evocative smell of woodsmoke now carries a very different meaning for her.

23.

We climb the broad, grassed flanks of Black
Knight Mountain, northeast of Kelowna. Artist–
naturalist Joanne Beaulieu walks ahead of me.
Recovering from cancer surgery, this is her first
time back on the grassland she knows and loves.
I expected to be accompanying her at a slow pace,
but instead we move along briskly; she has any
number of destinations she wants me to see. "I'll
probably pay dearly for this walk tonight, but I
don't care," she says, "It feels so wonderful to be
out here again."

I natter on to Joanne about the dilemma of
rich, fertile grasslands like this one, and how they
are far more prone to being hijacked by weeds
than grasslands on poor, bony soils. As I am talk-
ing, Joanne suddenly jerks her head upward and
clamps the binos to her eyes. When I first started

hanging out with birders, I thought these sudden conversational departures came from a combination of rudeness and attention deficit disorder. Now I realize these people have a special neural network, a kind of virtual cone that radiates from underneath their Tilley hats out about a kilometre. Every bird flying into that neural cone registers instantly. Like air traffic controllers, birders have to know the identity of every object on their screen. After a short, theatrical pause, Joanne downs her binos and closes the neural identity loop that links her to the black speck above us: "Swainson's hawk, no question. And look, there's its mate, off to the west." Joanne is definitely hardwired to birds.

We continue to ascend. The abundant dirt mounds of the pocket gophers provide me with a continuous readout of soil colour. I wish I had my Munsell Soil Colour Chart with me, as it splits ordinary brown into a dozen different gradients based on hue and chroma. At the base of the mountain, along the highway, the soil mounds are a pale beige, almost the colour of sandstone. As we move upward, the mounds take on a richer, more emphatically brown colour. At a certain point they morph into chestnut colour, until finally as we reach a mid-elevation bench, the mounds are the dark, rich colour of German chocolate cake. The colour of grassland soils is dictated by the amount of humus – decayed organic matter – produced by the death and decomposition of bunchgrass roots. Up here on the bench there is enough humus for the soil to qualify as a chernozem, that infinitely fertile, moisture-holding soil we have plowed up and converted to farmland right across the temperate grasslands of the planet. I kneel down and dig my hand into a particularly large gopher mound and rub the silky, almost greasy soil between my fingers. The deep colour stain stays, even after I wipe my fingers off on my trouser leg.

There is a small slough on the bench, ringed with aspen. Next to the slough is a broken-down corral. Cows and deer have eliminated all vegetation from ankle to about chest height. The tall saskatoon bushes are all mushroom-shaped, with bare, scarred stems supporting leafy growth that starts at the "ungulate line," just above the height that cows and deer can reach. The wretched state of the bench and its little slough should be cause for outrage, but with my ecological Pollyanna glasses, I see it transformed simply by managing the grazing and browsing.

Joanne and I have fallen into the typical naturalist hiking style of walking in separate but interwoven patterns. We each explore slightly different territories, but cross paths often to discuss findings. At one of our crossovers, she stops in mid-sentence to watch a pair of kestrels swoop gracefully over the grassland, but re-engages the sentence precisely, after the kestrels pass. This is further proof that I do not have what it takes to be a birder, since I can barely walk and chew gum at the same time. I tell Joanne about how the soil shift from brown to black mirrors a grass transition as well, from the tough and virtually drought-proof bluebunch wheatgrass to the more luxuriant Idaho fescue. Some of the dirt mounds are larger now and staffed with fat, inquisitive marmots. Joanne leads me to an even larger excavation and points out the formidable badger paw prints in the dirt.

As we continue to walk, I scan the ground for rattlesnakes, which have been sighted in the area. Essentially a desert animal, the rattlesnake ranges up the valley to about Vernon. These reptiles have always been part of my experience. Their enigmatic presence has been a constant from my youth onward, and if I put my mind to it, I could probably recall every single rattlesnake encounter. They are so completely alien to the human experience that I find them compelling. Joanne is hardwired to birds; I must be hardwired to rattlesnakes. Our Pacific rattlesnake has a very exacting set of habitat requirements. They need large, deeply fractured rock outcrops to hibernate in. Then, below the outcrop, they require a talus slope with rock fragments of a certain size so they can thermoregulate. Then, below the talus, there must be native grassland, but not just any grassland. The soils of the grassland must be loose, deep and favourable to the pocket gopher, the snakes' main food source. Needless to say, our Okanagan rattlesnake populations flirt with endangered status.

We finally stop to rest for a bit and Joanne describes her part-time job at Kelowna's nature interpretation centre, sited in remnant riparian forest along Mission Creek. She was opening the centre one bitterly cold December morning when a hooded and bundled-up vagrant, who had apparently spent the night outside, came to the door. Joanne suppressed her first instincts and let the fellow inside to warm up, and have a cup of coffee. Slowly the presumed street person unwrapped scarf, hood and toque to reveal the face of Dick Cannings, Canada's foremost ornithologist, who had spent the early part of the morning doing a Christmas bird count along the creek.

The endpoint of our journey is the summit of Black Knight Mountain, which presides over

Kelowna and Rutland. Like many mountains with commanding views, Black Knight once had a fire lookout tower. In the early days of firefighting, a ranger in the lookout tower would pinpoint the location of a new fire using a grid laid over high-quality landscape photographs taken from the tower. The ranger would radio the grid coordinates of the fire down to the base commander, who had an identical set of photos and could dispatch crews directly to the fire. This technology disappeared with the advent of airplane spotting, infrared scanning and satellite photography. Fortunately, the lookout photos were preserved. I managed to obtain a set of 1951 Black Knight tower photos from a colleague in the Forest Service. The tower had long since been torn down, but Joanne and I stood on the old footings and fiddled with camera angles and reference points until I could rephotograph the originals fairly precisely. With her artist's eye, she was able to interpret perspective and shadow in the original photographs, to help me make sense of them. I asked her if she had any insights for me about Black Knight Mountain. She thought for a minute and then said: "Don't forget the purples. You might think the shadows that fall on this mountain are all black, or grey. But within the shadows are wonderful blues, and shades of purple."

Don't forget the purples. I drove home to Summerland ruminating about connections between human health, the arts and nature.

24.

Gold is not a commodity we associate with the Okanagan, but in an indirect way, gold opened this valley. The fever that began with the 1849 California Gold Rush went viral. Mere rumors of promise in some distant creek or mountain were enough for men to pull up stakes and travel thousands of kilometres. Speed and early arrival were of the essence, as latecomers would find all the land already claimed and worked. The Cariboo Gold Rush in central British Columbia was such a spasm. Beginning around 1860, it fell between the California and the Klondike strikes, and attracted men from all over the continent. The primary route from the USA was up the Okanagan Valley, then through Kamloops to Clinton, and from there either north up to Barkerville or west to work claims along the Fraser River.

As a result of the Cariboo Rush, the very first Okanagan settler occupations were transportation, followed closely by government customs collection, and shortly after that, cattle ranching. Driving cattle all the way from the US up to the Cariboo to feed ravenous gold miners was a brutal and thankless task. Early on, several Okanagan entrepreneurs started ranches to gain a slice of the Cariboo provisioning market. These new ranchers were, almost to a man, either Scottish or Irish immigrants, often from wealthy families and typically ex-military. Men like Haynes, Allison, Ellis, Vernon and Dun-Waters established home ranches and set about amassing land and cattle. After the Gold Rush evaporated in 1865, the ranches continued and prospered. Three of the biggest early ranches bracketed the little community of Vernon: the O'Keefe to the north, the Coldstream to the east, and the Fintry to the south. The names of those pioneer ranch families now live on in the place names and local histories of the Okanagan.

Early life on the big Okanagan ranches was patterned after British sporting life, having much to do with horses, whisky, manorial homes and even fox hunts (coyotes were the unfortunate substitutes). The ranches took on more of a western flavour over time, but British Isles roots still remain strong in the Vernon and Armstrong area.

One hundred fifty years on, the proud Okanagan and Similkameen cattle ranching industry continues. Hayfields dot the valley bottoms, and summer pasture is found on Crown rangelands on both sides of the two valleys. A feedlot near Oliver helps round out the sector. But contemporary ranch operators face the mounting challenges of suburbanization, forest ingrowth, weed invasion and a gradual loss of the Okanagan's agricultural base. Ranchers and environmentalists, never great bedfellows, are beginning to find common cause in their mutual concern about the rapid expansion of suburbs into the grasslands.

The Okanagan–Similkameen did experience direct gold rush impact in the 1890s, with small strikes in Hedley, Olalla and Fairview. The villages of Hedley and Olalla survived when their respective mines petered out, but Fairview, west of Oliver, did not. It is instructive to walk the old Fairview townsite and to see the maps and old photos of the grid-patterned streets, the palatial hotel and the government offices. There is absolutely nothing left of Fairview now, except those old maps and photographs. Antelope brush grows where a town once stood. Nearly all BC communities were built around resource exploitation of one form or another, but Fairview is an extreme example of that ethos: take what you can as fast as you can, with no thought to a sustainable future.

25.

Heading northward from Kelowna, Highway 97 skirts Wood and Kalamalka Lakes, which lie parallel to Okanagan Lake but are separated from it by a narrow range of hills. Crossing in between the two lakes at Oyama, I visit a large private landholding on the east side of Kalamalka Lake, a lovely stretch of dry forest and grassland that the owner maintains as an ecological and wildlife reserve. As I walk the parcel, two disparate images float in my mind: the hourglass and the string of pearls. Let me explain.

Climate change will soon transform our Okanagan ecosystems. Scientists are noticing some climate-derived shifts already, but that steady trickle of change will inevitably swell. Species and ecosystems adapted to low elevation valley bottoms will see their climatic envelope move up the valley sides. Species whose ranges lie to the south of us will find that their favoured climatic niche has expanded northward. Nature shuns a vacuum; as native plants and animals vacate areas to which they are no longer adapted, a new set of plants and animals will take their place. This ecological re-sorting process can result in two vastly different outcomes. The newly vacated lands can be colonized by weedy, invasive alien species, or they can be colonized in an orderly progression of newly adapted *native* species moving northward and upward in elevation, in conveyor-belt fashion. The choice of outcomes has a lot to do with how we manage land, energy and expectations.

For my climate-change tripling, I chose a Gewürtztraminer from Salmon Arm. That area is currently the northernmost wine-growing area on the North American continent. Just a few decades ago, producing wine in Salmon Arm would have been unthinkable, partly because winter temperatures frequently dipped below that minus 23°c threshold. But those killing temperatures happen far less often now. In the future, we can expect vineyards to pop up in valleys even farther north than Salmon Arm. For the meal, I purchased some wild-caught sockeye salmon and broiled it gently with butter, lemon juice and dill. The choice of salmon was intentional, not because of Salmon Arm, but because this fish is iconic to British Columbia and climate change is adding lethally high river temperatures and low water flows to the salmon's existing stock of misery. I partook of the fine red flesh with a mixture of respect and remorse.

For the book, I chose a collection of essays, *Climate Change and Biodiversity*, edited by Lovejoy and Hannah. The 2005 book is already a classic in its field, documenting the many ways climate change will alter – and in some cases already has altered – biological life on the planet. As I scanned through Lovejoy and Hannah, one of the many new concepts in the book stood out: biological decoupling. To envision decoupling, think of a plant that relies on a specific pollinating insect, or a migratory bird that relies on certain seeds being available at a certain time. Then turn the climate knobs somewhat randomly, but mostly upward, and watch what happens. The new climate may suit the plant's physiology but not that of the pollinating insect. Or the migratory bird may arrive before the required seeds are ripe. A host of biological pairings, in the Okanagan Valley and beyond, can be decoupled by differential responses to climate change.

Now back to the images that were floating in my mind as I walked the shores of Kalamalka Lake. I like to think of the Okanagan Valley as the narrow neck of a continent-sized hourglass, with the Great Basin biome flaring to the south of us and the Thompson/Cariboo/Chilcotin country of the central interior swelling to the north. Great Basin flora and fauna are our nearest native neighbours to the south. They will be able to gradually migrate into the great new climate-altered spaces of British Columbia's central interior.

First, however, they will have to navigate through the fragmented, overdeveloped and exceedingly narrow neck of this hourglass: the Okanagan Valley. That navigation will be no mean feat, considering the amount of this valley that has been appropriated from nature for human use.

Species migrate at different rates. Flying insects can colonize a new site within a few days; trees may require decades or even centuries. Movement ability and means of transport are many and diverse. Some organisms can cross oceans, while others are stopped by a dirt road. This is where the string of pearls comes in.

Scattered through the Okanagan Valley are small tracts of park lands, ecological reserves, Important Bird Areas, conservation lands and private holdings like the one at Oyama. Together, they might form the beginnings of a more or less continuous string of healthy native habitats. A completed string of ecological pearls would allow the Great Basin species conveyor belt to dominate, instead of the alien invader conveyor belt.

We who live in and value the Okanagan have the collective responsibility to assemble, maintain and protect the string of pearls. Such an effort will require securement of a range of habitats, across a range of aspects, elevations, moisture gradients and geographic locations. At some point we may need to physically assist migration by hand carrying species or propagules to new locations, but that is a last resort. Nature does these things best; we just need to give her enough healthy land and water to do her rearranging in.

26.

The alternation of seasons is part of the rhythm of this valley. Each annual sequence brings new insights and new colours. A long, open fall such as this one turns the aspen leaves a yellow so intense that it must be proprietary. If paint manufacturers could reproduce it – which they can't – the colour would carry an upscale name like Aspen Elation. At this time of year, the groves of aspen seem to possess their own source of internal illumination. They bathe you in a quality of light worthy of the paintings of the Dutch masters. Looking up beyond the aspens of the valley bottom to the forested mountainsides in the distance, you see a tattered band of burnt orange amongst the green; that is the mid-elevation larches coming into their own particular fall colour. Above the larches, along the ridgelines, is a premonitory band of white. Soon the snowline will move downslope and gradually enclose the valley.

It is time to seal up the windows and fuss over the neglected woodpile. The chainsaw, the rototiller and the lawnmower all get a shot of fuel stabilizer. Faucets are drained and hoses put away. Days contract and horizons get shorter. A neglected pile of books and magazines next to the overstuffed chair suddenly become interesting. You stay pretty close to the woodstove, and notice the dog and cat are right there with you. The aquarium, mostly ignored all summer, comes into its own now. An old fleece vest becomes part of your daily fashion statement. You remember the modest delights of winter vegetables like carrots, cabbage, rutabaga. You step back for a moment and celebrate the functionality of your house. It is then you realize you are participating once again in the universal pan-Canadian event: winter.

Home and place are elusive. I have been chasing them most of my life through thickets of ambiguity. Their paths cross, only to diverge again.

Yet, as the Okanagan's landscapes and seasons glide by me, the two concepts come ever closer to my grasp. I am beginning to understand that for a house to be a home, it must be situated in a regional place you know. Some say the geography of one's place should be bioregional; let watersheds define it. As I mentally scan the Okanagan watershed, I think that concept works fine for me. The Okanagan encompasses an entire world of people, place, water and possibility. Yet, in my investigations I can traverse it in just a few hours.

Our English parlance allows me to use the word *home* to refer ambiguously to my house, my town or my region. I can also refer to my dwelling as either *my home* or *my place*. There is intent behind those semantic confusions; they did not happen by accident.

Often when I tell people I live in the Okanagan, their response is, "Oh, you're so lucky, I wish I could live there." Curiously, I don't like hearing that. I don't like placing one bioregion above another. I also don't like the idea that by simply moving to a place, we gain some of the cachet of that place. Mere association with a place does not transform our lives. I don't want my adopted home region to be seen as something to be captured or owned. The Okanagan is not a trophy wife.

This odyssey has helped my placemaking. From the sagebrush of Chopaka to the opposing waters of Armstrong, I have been able to put flesh on my valley. Now I shall move on to further triplings of locally meaningful books, wine and food, against a backdrop of local landscapes. This is my commitment: to honour the Okanagan terroir.

BIBLIOGRAPHY

Association of British Columbia Grape Growers. *Atlas of Suitable Grape Growing Locations in the Okanagan and Similkameen Valleys of British Columbia.* Summerland, BC: Soil Science & Agricultural Engineering, Agriculture Canada, 1984.

Brautigan, Richard. *A Confederate General From Big Sur.* New York: Grove Press, 1970.

Haymour, Eddy. *From Nut House to Castle: The Eddy Haymour Story.* Peachland, BC: Stone Publishing, 1992.

Hessing, Melody. *Up Chute Creek: An Okanagan Idyll.* Kelowna, BC: Okanagan Institute, 2009.

Jacobs, Jane. *The Death and Life of Great American Cities.* New York: Random House/ Vintage, 1961/1992.

Keast Lord, John. *The Naturalist in Vancouver Island and British Columbia,* vol. 2. London: R. Bentley, 1866. Available online at http://is.gd/aTBfQ.

Lazarus, John. *The Trials of Eddy Haymour.* Toronto: Playwrights Guild of Canada, 1997.

Lea, Ted. "Historical (pre-settlement) Ecosystems of the Okanagan Valley and Lower Similkameen Valley of British Columbia, pre-European Contact to the Present." *Davidsonia* 19, no. 1 (2008): 3-36. Available online at www.davidsonia.org/files/david sonia_19_1.pdf

Lovejoy, Thomas E., and Lee Jay Hannah. *Climate Change and Biodiversity.* New Haven, Conn.: Yale University Press, 2005.

Marshall, James. *Let's Not Let a Little Thing Like the End of the World Come between Us.* Saskatoon: Thistledown Press, 2004.

Pyne, Stephen J. *Fire: A Brief History.* Seattle: University of Washington Press, 2001.

Rhenisch, Harold. *Out of the Interior: the Lost Country.* Vancouver: Ronsdale Press, 1994.

Ryga, George. *Summerland.* Vancouver: Talonbooks, 1992.

White, Patrick. *The Tree of Man.* New York: Viking Press, 1955.

Selected General References on the Okanagan/Similkameen

Cannings, Richard. *Roadside Nature Tours through the Okanagan.* Vancouver: Greystone Books, 2009.

Koroscil, Paul M. *The British Garden of Eden: Settlement History of the Okanagan Valley, British Columbia.* Burnaby, BC: Simon Fraser University Dept. of Geography, 2008.

Parish, Roberta, R. Coupe, Dennis Lloyd and Joe Antos. *Plants of Southern Interior British Columbia.* Vancouver: Lone Pine, 1996.

Allison, Susan. *A Pioneer Gentlewoman in British Columbia: The Recollections of Susan Allison.* Edited by Margaret A. Ormsby. Recollections of the Pioneers of British Columbia, vol. 2. Vancouver: University of British Columbia Press, 1976.

Sanford, Barrie. *McCulloch's Wonder: The Story of the Kettle Valley Railway.* West Vancouver, BC: Whitecap Books, 1977.